ON SOME ASPECTS OF HEIGHT CONVERSION AND VERTICAL DATUM UNIFICATION

高程系统转换及垂直基准统一的若干问题研究

Robert Tenzer 著

WUHAN UNIVERSITY PRESS
武汉大学出版社

图书在版编目(CIP)数据

高程系统转换及垂直基准统一的若干问题研究 = On Some Aspects Of
Height Conversion And Vertical Datum Unification:英文/(斯洛伐)藤泽尔
(Tenzer,R.)著.—武汉:武汉大学出版社,2015.7
　　ISBN 978-7-307-15883-2

　　Ⅰ.高…　　Ⅱ.藤…　　Ⅲ.①高程系统—研究—英文　　②大地测量基
准—研究—英文　　Ⅳ.①P216　②P22

　　中国版本图书馆 CIP 数据核字(2015)第 110623 号

责任编辑:谢群英　　　　责任校对:汪欣怡　　　　版式设计:马　佳

出版发行:**武汉大学出版社**　　(430072　武昌　珞珈山)
　　　　　(电子邮件:cbs22@whu.edu.cn　网址:www.wdp.com.cn)
印刷:武汉中远印务有限公司
开本:720×1000　　1/16　　印张:8.5　　字数:120 千字　　插页:1
版次:2015 年 7 月第 1 版　　2015 年 7 月第 1 次印刷
ISBN 978-7-307-15883-2　　　　定价:26.00 元

Preface

The research described herein was conducted during my postdoctoral stay in the Department of Geomatics at the University of New Brunswick in Canada and later continued during my lecturing in the National School of Surveying at the University of Otago in New Zealand and in the School of Geodesy and Geomatics at the Wuhan University in China. Theoretical definitions of the rigorous orthometric height and the geoid-to-quasigeoid correction were discussed with Prof. Petr Vaníček (University of New Brunswick), Prof. Will E. Featherstone (Curtin University), Prof. Lars E. Sjöberg (Royal Institute of Technology), Prof. Pavel Novák (University of West Bohemia), Dr. Christian Hirt (Curtin University), and Dr. Sten Claessens (Curtin University). The presented numerical results were compiled with the help of Dr. Nadim Dayoub (University of Newcastle upon Tyne), Dr. Robert Čunderlík (Slovak Technical University), Prof. Viliam Vatrt (Brno University of Technology) and my former PhD student Dr. Ahmed Abdalla (University of Otago). The digital density model of New Zealand was complied with the help of Dr. Pascal Sirguey (University of Otago) and the advice of Dr. Mark Rattenbury (GNS Science). The gravity database was kindly provided by the GNS Science New Zealand, and the levelling and GPS data and the NZGeoid2009 official quasigeoid model of New Zealand by the Land Information New Zealand.

The publication of this monograph was supported by the National Science Foundation of China (NSFC), grant no. : 41429401.

Summary

The definition and practical realization of the World Height System (WHS) requires the unification of several continental, national and local geodetic vertical controls currently established over the world. This can be done partially (on a continental scale) by a joint adjustment of levelling networks, while a global realization requires finding their relation with respect to the geoidal geopotential value W_0. Another major issue, associated with the vertical datum unification, is a choice of a height system. Either Helmert's orthometric heights or Molodensky's normal heights are practically used in countries where the leveiling networks were realized through geodetic spirit levelling and gravity measurements along levelling lines. In countries, where these gravity measurements are absent, the normal gravity values were used to approximate the actual gravity. The vertical datum is in this case defined in the system of normal-orthometric heights. The conversion between different types of heights is thus indispensible for the unification of geodetic vertical datums. The rigorous relation between the orthometric and normal heights is utilized in definitions of the geoid-to-quasigeoid correction in the spatial and spectral domains presented in this work. However, the practical application of these expressions in computing the geoid-to-quasigeoid correction is not simple, because it requires the knowledge of the terrain geometry, topographic density distribution and importantly also the facilitation of advanced

1

numerical techniques. After reviewing the numerical models of computing the geoid-to-quasigeoid correction in the spatial and spectral domains, some of these practical aspects are discussed in the context of the (experimental) vertical datum unification in New Zealand. The height reference system in New Zealand was realized by several local vertical datums (LVDs), which were established throughout the country based on precise levelling from tide gauges or connecting to existing levelling networks. Moreover, the LVDs were defined in the system of the normal-orthometric heights, because of the absence of measured gravity values along levelling lines. Asserting that all geodetic data available should be incorporated in the vertical datum realization, the unification of LVDs at the North and South Islands of New Zealand was realized in several processing steps, which comprised the levelling network adjustment, the gravimetric geoid and quasigeoid determination, the height conversion, the compilation of digital terrain and density models, the analysis of the mean dynamic topography offshore, the conversion between permanent tidal systems, and the estimations of LVD offsets. These procedures are summarized here and possible methods of improving the accuracy of the height conversion are also discussed.

A Table of Contents

1. Introduction

For a practical realization of the geodetic vertical datum, Helmert's (1884, 1890) orthometric heights are preferably used. The reason is a simple computation of the mean gravity using Poincaré-Prey's gravity reduction while assuming a uniform topographic density distribution, and the acceptable accuracy for most of the regions where the levelling networks are established. To determine the orthometric heights in the mountainous, polar and geologically complex regions with the accuracy of a few centimetres or even better, Helmert's definition is not sufficient. In this case, more accurate methods for the evaluation of mean gravity have to be applied.

A more accurate method was introduced by Niethammer (1932, 1939). He took the terrain effect into consideration while assuming a uniform topographic density. According to his method, the mean value of the planar terrain correction is evaluated as a simple average of values computed at the finite number of points along the plumbline within the topography; see also Baeschlin (1948) who summarized his work. Mader (1954) estimated the difference between Helmert and Niethammer's methods of ~6 cm for Hochtor (2,504 m) in the European Alps, see also Heiskanen and Moritz (1967, Chapters 4-6). Mader (1954) and Ledersteger (1968) also presupposed that the terrain correction varies linearly with depth. Based on this assumption, the mean terrain correction is averaged from two values computed

1

for points at the topographic surface and the geoid. Flury and Rummel (2009), however, demonstrated that the non-linear changes of the terrain correction could not be disregarded. Hence, the mean terrain correction should be computed according to Niethammer (1939) or Flury and Rummel (2009). Wirth (1990) modified Niethammer's method by means of computing the topographic gravity potential (instead of the terrain correction) at points at the topographic surface and the geoid.

It is a well-known fact that the mean gravity within the topography depends also on the actual topographic density distribution. The variation of topographic density can cause changes in orthometric height up to several decimetres (e. g. Vaníček et al. , 1995). The correction to Helmert's orthometric height due to the lateral variation of topographic density can be evaluated using a simple formula in which the change of orthometric height is in a linear relation to the anomalous lateral topographic density (Heiskanen and Moritz, 1967). Adopting this relation, the effect of the anomalous topographic density to Helmert's orthometric heights was investigated, for instance, by Allister and Featherstone (2001) and Tenzer and Vaníček (2003).

In these approximate definitions of the orthometric height, the vertical gravity gradient generated by the mass density distribution below the geoid surface is approximated by the linear normal gravity gradient while disregarding the change of the normal gravity gradient with depth. Hwang and Hsiao (2003) estimated that this approximation causes the inaccuracy of orthometric heights up to several centimetres in the mountainous regions.

Tenzer and Vaníček (2003) applied the analytical downward continuation of the observed gravity in the evaluation of the mean gravity along the plumbline within the topography based on assuming the lateral topographic density distribution. They then formulated the relation between Poincaré-Prey's gravity gradient and the analytical downward continuation of gravity. A more accurate method for a determination of the mean gravity was introduced by Tenzer et al. (2005) . They applied the decomposition of the mean gravity into the mean normal gravity, the mean no-topography gravity disturbance (generated by the mass density distribution below the geoid surface) and the mean values of the gravitational attractions of topographic and atmospheric masses. The mean normal gravity is evaluated according to Somigliana-Pizzetti's theory of the normal gravity field (Pizzetti, 1911; Somigliana, 1929). The mean topography-generated gravitational attraction is, in accordance with Bruns' (1878) theorem, defined in terms of the difference of gravitational potentials reckoned to the geoid and the topographic surface, multiplied by the reciprocal value of the orthometric height. The same principle was deduced for a definition of the mean atmosphere-generated gravitational attraction. The mean no-topography gravity disturbance is defined by applying Poisson's integral to the integral mean and solving the inverse to Dirichlet's boundary-value problem for the downward continuation of the no-topography gravity disturbances in prior of computing the integral mean value. In addition to the above theoretical developments, numerous empirical studies have been published on the orthometric height definition (e. g. , Ledersteger, 1955; Rapp, 1961; Krakiwsky, 1965; Strange, 1982; Sünkel, 1986; Kao et al. , 2000; Tenzer and Vaníček, 2003; Dennis and Featherstone, 2003).

Asserting that the topographic density and the actual vertical gravity gradient inside the Earth could not be determined precisely, Molodensky (1945, 1948) formulated the theory of normal heights. In his definition, the mean actual gravity within the topography is replaced by the mean normal gravity between the reference ellipsoid and telluroid (see also Heiskanen and Moritz, 1967, Chapter 4). The normal heights are thus defined without any hypothesis about the topographic mass density distribution.

The Molodensky normal heights and the Helmert orthometric heights are the most widely-used height systems. These two types of heights can be adopted if the levelling networks were established based on geodetic spirit levelling and gravity measurements along levelling lines. In some countries, however, the gravity values along levelling lines were calculated only approximately using the normal gravity. The vertical datum is then defined by the normal-orthometric heights.

In recent years, a considerable effort has been undertaken to unify a large number of existing vertical datum realizations around the world. The vertical datum unification typically requires the joint adjustment of interconnected levelling networks and/or the definition of the vertical datum offset with respect to the World Height System (WHS), which is defined by the geoidal geopotential value W_0 . Alternatively, the vertical datum unification can be realized though the gravimetric determination of the global geoid/quasigeoid model to a high accuracy and resolution. Since the geodetic vertical systems are defined using different types of heights (and every country adopted their own height system specifications), the conversion between these types of heights is inevitable. The height conversion has been addressed extensively in geodetic literature. An approximate formula relating the normal and orthometric heights was given, for instance, in Heiskanen and Moritz (1967, Eqs. 8-103). Sjöberg (1995) slightly improved the classical definition by adding a small correction term related with the vertical derivative of the gravity anomaly. Tenzer et al. (2005) presented numerical procedures for a rigorous computation of the orthometric height and formulated an accurate relation between the (rigorous) orthometric and normal heights. An alternative method of computing the geoid-to-quasigeoid correction was given by Tenzer et al. (2006). They derived this correction based on comparing the geoidal height and the height anomaly, both defined by means of applying Bruns' (1878) theorem. A very similar expression for computing the geoid-to-quasigeoid correction was given by Sjöberg (2006). The definitions of the geoid-to-quasigeoid correction presented by Tenzer et al. (2005, 2006) and Sjöberg (2006) incorporated information on the terrain geometry, variable topographic density and mass density heterogeneities distributed below the geoid surface. Santos et al. (2006) investigated the relations between various types of the orthometric height definitions. Flury and Rummel (2009) investigated the effect of terrain geometry to the geoid-to-quasigeoid correction. They demonstrated that the consideration of the terrain geometry significantly reduces the values of the geoid-to-quasigeoid correction computed using the classical definition in which the topography is approximated by the Bouguer plate. The results of Flury and Rummel (2009) were in a good agreement with previous results over larger area in European Alps presented by Marti (2005) and Sünkel et al. (1987) (see also Hofmann-Wellenhof and Moritz, 2005) Following the work of Flury and Rummel (2009), Sjöberg (2010) derived a slightly more accurate expression for the geoid-to-quasigeoid correction, consistent with a definition of the boundary condition of physical geodesy (see also Sjöberg and Bagherbandi, 2012; Bagherbandi and Tenzer, 2013). He, however, also stated that his more refined expression could improve the accuracy not more than ~1 cm compared to the expression given by Flury and Rummel (2009). Later, Sjöberg (2012) applied an arbitrary compensation model in computing the topographic correction term. In particular, he recommended using either the Helmert or isostatic types of reductions, which provide smaller and smoother components, more suitable for interpolation and calculation, than the Bouguer reduction. It is worth mentioning herein that the conversion of the normal-orthometric to normal heights was applied, for instance, by Filmer et al. (2010) and Tenzer et al. (2011a, 2011b).

To begin with, the fundamental definitions in the theory of heights are here briefly recapitulated. With reference to these definitions, the expressions for an accurate conversion between the normal and orthometric heights (i. e. , the geoid-to-quasigeoid correction) are then presented in the spatial and spectral domains. The numerical procedures of computing the geoid-to-quasigeoid correction in the spatial domain are compared. The computation of this correction in the spectral domain is realized by means of applying methods for a spherical harmonic analysis and synthesis of the gravity field and continental crustal density structures. The geoid-to-quasigeoid correction could be computed accurately only if the actual crustal density distribution within the topography is known to a sufficient accuracy. Moreover, this computation utilizes relatively complex numerical schemes which cannot routinely be applied in practice. Therefore, the rigorous definition of the orthometric height (and consequently the geoid-to-quasigeoid correction) is likely to be restricted mainly to scientific purposes, while its use in broader, more practical geodetic applications remains limited. Possible reasons are discussed in the context of (experimental) vertical datum

6

unification in New Zealand, conducted in several processing steps. These steps comprise the levelling network adjustment, the gravimetric geoid and quasigeoid modelling, the estimation of LVD offsets, the conversion between the permanent tidal systems, the analysis of systematic errors, and the conversion between different height systems. Moreover, in order to improve the accuracy of computing the geoid-to-quasigeoid correction, digital terrain and density models are needed. For this purpose, the rock density model was compiled from existing geological maps, rock density samples, and additional geological sources. This model is then facilitated in the gravimetric forward modelling of variable topographic density.

The content is organized into eight chapters. The following three chapters provide a brief summary of the coordinate systems and transformations (Chapter 2), the Earth's gravity field (Chapter 3) and the theory of heights (Chapter 4). The explicit definition of the geoid-to-quasigeoid correction and the expressions used for computing this correction in the spatial and spectral domains are given in Chapter 5. The practical aspects of vertical datum unification in New Zealand are discussed, and numerical results presented in Chapter 6. The effect of variable topographic density on gravity field quantities is investigated in Chapter 7. The summary and major conclusions are given in Chapter 8.

2. Coordinate Systems and Transformations

The 3-D position is defined in the Cartesian coordinate system (X, Y, Z) of which the origin is identical to the mass center of the Earth, the Z-axis pass through the Conventional International Origin (CIO), and the X-axis pass through the intersection of the Greenwich meridian plane with the equatorial plane. Analogously, the 3-D position can be described by the geodetic coordinates (h, φ, λ) or the spherical coordinates (r, ϕ, λ), where φ and λ are the geodetic latitude and longitude respectively, ϕ is the spherical latitude, the spherical and geodetic longitudes λ are identical, r is the geocentric radius, and h is the geodetic (ellipsoidal) height (see Fig. 2. 1).

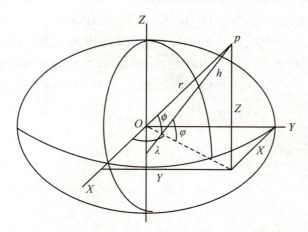

Fig. 2. 1　The Cartesian, geodetic and spherical coordinate systems

The direct transformation between the Cartesian and geodetic coordinate systems is defined by (e. g. , Bomford, 1971)

$$X = (R_N + h)\cos\varphi\cos\lambda ,Y = (R_N + h)\cos\varphi\sin\lambda ,$$
$$Z = [R_N(1 - e^2) + h] \sin\varphi . \qquad (2.1)$$

The inverse transformation can be realized iteratively, or directly using the following expressions:

$$\tan\varphi = \frac{Z + be'^2 \sin^3\beta}{\sqrt{X^2 + Y^2} - ae^2 \cos^3\beta} , \quad \tan\lambda = \frac{Y}{X} ,$$

$$h = \frac{\sqrt{X^2 + Y^2}}{\cos\varphi} - R_N, \quad \tan\beta = \frac{Z}{\sqrt{X^2 + Y^2}} \frac{a}{b} , \qquad (2.2)$$

where the prime-vertical radius of curvature R_N reads

$$R_N = \frac{a}{\sqrt{1 - e^2 \sin^2\varphi}} . \qquad (2.3)$$

The linear flattening f, the first e and second e' numerical eccentricities of the reference ellipsoid are defined as follows (e. g. , Bomford, 1971)

$$f = \frac{a - b}{a} = 1 - \frac{b}{a} = 1 - \sqrt{1 - e^2} = 1 - \frac{1}{\sqrt{1 + e'^2}} , \qquad (2.4)$$

$$e^2 = \frac{a^2 - b^2}{a^2} = 2f - f^2 = 1 - \frac{b^2}{a^2} = 1 - \frac{1}{1 + e'^2} = \frac{e'^2}{1 + e'^2} , (2.5)$$

$$e'^2 = \frac{a^2 - b^2}{b^2} = \frac{2f - f^2}{(1 - f)^2} = \frac{1}{1 - e^2} - 1 = \frac{e^2}{1 - e^2} . \qquad (2.6)$$

The transformation between the Cartesian and spherical coordinate systems is given by(ibid.)

$$X = r\cos\phi\cos\lambda , \quad Y = r\cos\phi\sin\lambda , \quad Z = r\sin\phi , \qquad (2.7)$$

and its inverse formulas read as

$$\tan\phi = \frac{Z}{\sqrt{X^2 + Y^2}}, \quad \tan\lambda = \frac{Y}{X}, \quad r^2 = X^2 + Y^2 + Z^2 . \qquad (2.8)$$

9

3. Earth's Gravity Field

The disturbing potential T at an arbitrary point (r,Ω) is defined by (e. g. , Heiskanen and Moritz, 1967)

$$T(r,\Omega) = W(r,\Omega) - U(r,\phi) , \qquad (3.1)$$

where W is the Earth's gravity potential, and U is the normal gravity potential. The 3-D position is defined in the geocentric spherical coordinates (r,Ω) , where r is the geocentric radius, and the spherical direction is denoted as $\Omega = (\phi,\lambda)$.

By analogy with Eq. (3.1), the gravity disturbance δg at a point (r,Ω) is defined as the difference between the Earth's gravity g and the normal gravity γ . Hence

$$\delta g(r,\Omega) = g(r,\Omega) - \gamma(r,\phi) . \qquad (3.2)$$

Adopting the spherical approximation, the relations between the potential and gravity field quantities in Eqs. (3.1) and (3.2) are defined approximately by

$$\delta g \cong -\frac{\partial T}{\partial r} , \quad g \cong -\frac{\partial W}{\partial r} , \quad \gamma \cong -\frac{\partial U}{\partial r} . \qquad (3.3)$$

The gravity anomaly Δg is defined by the fundamental formula of physical geodesy. In the spherical approximation, it reads (e. g. , Heiskanen and Moritz, 1967)

$$\Delta g(r,\Omega) \cong -\frac{\partial T(r,\Omega)}{\partial r} - \frac{2}{r}T(r,\Omega) . \qquad (3.4)$$

The disturbing potential in Eq. (3.1) can be described in terms of the spherical harmonics (e. g. , Heiskanen and Moritz, 1967)

$$T(r,\Omega) = \frac{GM}{R} \sum_{n=0}^{\infty} \sum_{m=-n}^{n} \left(\frac{R}{r}\right)^{n+1} T_{n,m} Y_{n,m}(\Omega) \ , \qquad (3.5)$$

where $GM = 3,986,005 \times 10^8$ m^3 s^{-2} is the geocentric gravitational constant, $R = 6,371 \times 10^3$ m is the Earth's mean radius (which approximates the geocentric radii of the geoid surface), $Y_{n,m}$ are the (fully-normalized) surface spherical functions of degree n and order m, and $T_{n,m}$ are the (fully-normalized) numerical coefficients which describe the disturbing potential T. The coefficients $T_{n,m}$ are obtained from the coefficients of a global gravitational model (GGM) after subtracting the spherical harmonic coefficients of the GRS-80 normal gravity field (Moritz, 2000). Note that different values can be used to define parameters GM and R.

From Eq. (3.5), the gravity disturbance δg is defined in terms of spherical harmonics by the following expression

$$\delta g(r,\Omega) = \frac{GM}{R^2} \sum_{n=0}^{\infty} \sum_{m=-n}^{n} \left(\frac{R}{r}\right)^{n+2} (n+1) T_{n,m} Y_{n,m}(\Omega) \ . \qquad (3.6)$$

Similarly, the gravity anomaly Δg is given by

$$\Delta g(r,\Omega) = \frac{GM}{R^2} \sum_{n=0}^{\infty} \sum_{m=-n}^{n} \left(\frac{R}{r}\right)^{n+2} (n-1) T_{n,m} Y_{n,m}(\Omega) \ . \qquad (3.7)$$

4. Theory of Heights

The orthometric height H^o is defined as the distance, measured positive outwards along the plumbline, from the geoid (zero orthometric height) to a point of interest, usually at the topographic surface. The curved plumbline is at every point tangential to the gravity vector generated by the Earth, its atmosphere and rotation. The orthometric height can be computed from the potential number if available, using the mean gravity along the plumbline between the geoid and topographic surface. Alternatively and most commonly used in practice, it can be computed from spirit levelling measurements using the orthometric correction, embedded in which is the mean value of gravity (cf. Strang van Hees, 1992). Ignoring levelling errors and many issues associated with a practical vertical datum realization (e. g. , Drewes at al. , 2002; Lilje, 1999), the rigorous determination of the orthometric height reduces to an accurate determination of the mean gravity along the plumbline within the topography. The most generalized definition of the orthometric height $H^o(\Omega)$ is given by (e. g. , Heiskanen and Moritz, 1967, Eg. 4-21)

$$H^o(\Omega) = \frac{C(r_t,\Omega)}{\bar{g}(\Omega)} , \qquad (4.1)$$

where C is the potential number of a point at the topographic surface (r_t,Ω) of which the geocentric radius is denoted as r_t. The mean actual gravity along the plumbline between the

topographic surface and geoid is denoted as $\bar{g}$. The potential
number C is computed from measured height differences ΔH_i and
observed gravity values g_i along the levelling line as $C = \sum_i g_i \Delta H_i$.

The mean gravity $\bar{g}$ along the plumbline in Eq. (4.1) is defined as
follows (Tenzer, 2004)

$$\bar{g}(\Omega) = \frac{1}{H^0(\Omega)} \int_{r=r_g}^{r_t} g(r,\Omega) \cos(-g,r^o) dr , \qquad (4.2)$$

where $\cos(-g, r^o)$ is the cosine of the deflection of the
plumbline from the geocentric radial direction, g is the vector of
gravity, r^o is the unit vector in the geocentric radial direction, and
the geocentric radius of the geoid surface is denoted as r_g . Tenzer
et al. (2005) demonstrated that neglecting the deflection of the
plumbline from the geocentric radial direction could yield the
inaccuracy of orthometric height up to a few millimetres.

An appropriate method for the evaluation of the mean gravity
has been discussed for more than a century. The first theoretical
attempt is attributed to Helmert (1884, 1890). In Helmert's
definition of the orthometric height, the Poincaré-Prey gravity
gradient is used to evaluate the approximate value of mean gravity
$\tilde{g}^H$ from the observed gravity g at the topographic surface (also
see Heiskanen and Moritz, 1967, Chapter 4).

$$\tilde{g}^H(\Omega) \cong g(r_t,\Omega) - \left.\frac{\partial g(r,\Omega)}{\partial H}\right|_{r=r_t} \frac{H^0(\Omega)}{2} \cong$$

$$\approx g(r_t,\Omega) - \left(\frac{\partial\gamma(r,\phi)}{\partial h} + 4\pi G\rho^T\right)\frac{H^0(\Omega)}{2} , \qquad (4.3)$$

where $G = 6.674 \times 10^{-11}$ m^3 kg^{-1} s^{-2} is Newton's gravitational
constant, ρ^T is the average topographic density, $\partial g/\partial H$ and $\partial\gamma/\partial h$
are the actual and normal linear gravity gradients respectively.
According to Eq. (4.3), the approximate value $\tilde{g}^H$ is evaluated

13

so that the observed gravity at the topographic surface is reduced to the mid-point of the plumbline $H^o/2$ by applying the normal linear gravity gradient and assuming a uniform topographic density distribution.

Niethammer (1932) computed the mean gravity along the plumbline $\tilde{g}^N$ as follows

$$\tilde{g}^N(\Omega) \cong \tilde{g}^H(\Omega) - g^{TC}(r_t,\Omega) + \overline{g}^{TC}(\Omega) , \qquad (4.4)$$

where the planar terrain correction g^{TC} computed at the topographic surface is subtracted from the value of $\tilde{g}^H$ and the mean value of the planar terrain correction $\overline{g}^{TC}$ is added instead. This mean value is evaluated as a simple average of values computed at a finite number of points along the plumbline within the topography.

Mader (1954) evaluated (less accurately) the mean terrain correction along the plumbline as an average of two values computed for points at the topographic surface and the geoid using the following formula

$$\tilde{g}^M(\Omega) \cong \tilde{g}^H(\Omega) - \frac{g^{TC}(r_t,\Omega) + g^{TC}(r_g,\Omega)}{2} . \qquad (4.5)$$

Asserting that the topographic density and the actual vertical gravity gradient inside the Earth could not be determined precisely, Molodensky (1945, 1948) formulated the theory of normal heights H^N (see also Molodensky et al. , 1960). In his definition, the mean actual gravity within the topography is replaced by the mean normal gravity $\overline{\gamma}$ between the reference ellipsoid and telluroid. Hence

$$H^N(\Omega) = \frac{C(r_t,\Omega)}{\overline{\gamma}(\Omega)} , \qquad (4.6)$$

where the mean normal gravity is evaluated according to

Somigliana-Pizzetti's theory of the normal gravity field (Pizzetti, 1911; Somigliana, 1929).

In some countries, the gravity measurements were not conducted along levelling lines. In this case, the gravity values are computed approximately using the normal gravity. The vertical datum is then defined by the normal-orthometric heights (e. g. , Heiskanen and Moritz, 1967)

$$H^{N-O}(\Omega) = \frac{C^N(r_t, \Omega)}{\overline{\gamma}(\Omega)} , \qquad (4.7)$$

where the normal potential number C^N is computed from measured height differences ΔH_i and computed values of the normal gravity γ_i along levelling lines, i. e. , $C^N = \sum_i \gamma_i \Delta H_i$.

For some specific applications, such as the levelling network adjustment, the dynamic heights can be used. The dynamic height H^D is defined as (e. g. , Heiskanen and Moritz, 1967)

$$H^D(\Omega) = \frac{C(r_t, \Omega)}{\gamma_{0, \varphi = \text{const}}} , \qquad (4.8)$$

where the constant value of the normal gravity $\gamma_{0, \varphi = \text{const}}$ is conveniently computed at the ellipsoid surface for a chosen constant value of the geodetic latitude $\varphi = \text{const}$.

As illustrated in Fig. 4. 1, the geodetic height h is related to the orthometric H^O and normal H^N heights as follows

$$h = H^O + N = H^N + \zeta , \qquad (4.9)$$

where ζ is the height anomaly, and N is the geoid height. Sjöberg (2006) estimated that the deflection of the plumbline could cause maximum differences between the terms of $H^O + N$ and $H^N + \zeta$ to ~1. 5 mm.

In practical applications, such as the global geopotential model testing or the vertical datum unification, the problem of height conversion is almost exclusively related to a determination

15

of the geoid-to-quasigeoid correction. This correction is then applied to the normal heights in order to obtain the orthometric heights or vice versa. Several authors derived the expressions for computing the geoid-to-quasigeoid correction either in the closed analytical or spectral forms. For most of the practical purposes, this correction can be computed approximately as the difference between the Helmert orthometric height and the Molodensky normal height. From Eqs. (4. 1) and (4. 6), we get

$$H^N(\Omega) - H^O(\Omega) = \frac{H^N(\Omega)}{\bar{g}(\Omega)} [\bar{g}(\Omega) - \bar{\gamma}(\Omega)] \ . \qquad (4.10)$$

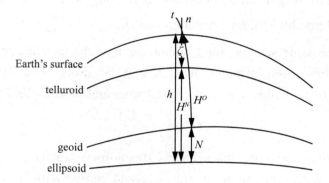

Fig. 4. 1 *The geodetic, orthometric and normal heights.*

We further define the mean normal gravity $\bar{\gamma}$ in Eq. (4. 10) by means of the normal linear gravity gradient

$$\bar{\gamma}(\Omega) \approx \gamma_0(\phi) + \frac{\partial \gamma(r,\phi)}{\partial h} \frac{H^N(\Omega)}{2} \ , \qquad (4.11)$$

where γ_0 is the normal gravity at the ellipsoid surface. Inserting from Eqs. (4. 3) and (4. 11) to Eq. (4. 10), we arrive at

$$H^N(\Omega) - H^O(\Omega) \cong \frac{H^N(\Omega)}{\bar{g}(\Omega)} \left[g(r_t,\Omega) - \frac{\partial \gamma(r,\phi)}{\partial h} \frac{H^O(\Omega)}{2} \right.$$

$$- 2\pi G \rho^T H^O(\Omega) - \gamma_0(\phi) - \frac{\partial \gamma(r,\phi)}{\partial h} \frac{H^N(\Omega)}{2} \Bigg] .$$

$$\tag{4.12}$$

Since

$$\frac{\partial \gamma(r,\phi)}{\partial h} \frac{H^O(\Omega)}{2} + \frac{\partial \gamma(r,\phi)}{\partial h} \frac{H^N(\Omega)}{2} \approx \frac{\partial \gamma(r,\phi)}{\partial h} H(\Omega) ,$$

$$\tag{4.13}$$

we can write (e. g. , Heiskanen and Moritz, 1967)

$$H^N(\Omega) - H^O(\Omega) \cong \frac{H}{\bar{g}(\Omega)} \Bigg[g(r_t,\Omega) - \gamma_0(\phi)$$

$$- \frac{\partial \gamma(r,\phi)}{\partial h} H(\Omega) - 2\pi G \rho^T H(\Omega) \Bigg]$$

$$\cong \frac{H(\Omega)}{\bar{\gamma}(\Omega)} \Delta g^{SPB}(r_t,\Omega) , \tag{4.14}$$

where H is the topographic height. The simple planar Bouguer gravity anomaly Δg^{SPB} in Eq. (4.14) is computed from the observed gravity anomaly Δg by subtracting the simple planar Bouguer reduction

$$\Delta g^{SPB}(r_t,\Omega) = \Delta g(r_t,\Omega) - 2\pi G \rho^T H(\Omega) . \tag{4.15}$$

Flury and Rummel (2009) improved the approximate formula in Eq. (4.14) by applying the additional correction terms which takes into consideration the topographic potential difference at the topographic surface and the geoid, i. e. ,

$$H^N(\Omega) - H^O(\Omega)$$

$$\cong \frac{H(\Omega)}{\bar{\gamma}(\Omega)} \Delta g^{RPB}(r_t,\Omega) + \frac{V^T(r_g,\Omega) - V^T(r_t,\Omega)}{\bar{\gamma}(\Omega)}$$

$$+ \frac{\overline{\delta g}^{RPB}(\Omega) - \delta g^{RPB}(r_t,\Omega)}{\bar{\gamma}(\Omega)} H(\Omega) , \tag{4.16}$$

where the values of the topographic potential V^T are computed at the geoid and the topographic surface, $\Delta g^{RPB}(r_t,\Omega) = \Delta g(r_t,\Omega)$ $- 2\pi G\rho^T H^O(\Omega) + g^{TC}$ is the refined planar Bouguer gravity anomaly which comprises the planar terrain correction term g^{TC}. The last constituent in Eq. (4. 16) defines the contribution of mass density heterogeneities below the geoid surface, where δg^{RPB} and $\overline{\delta g}^{RPB}$ is the Bouguer gravity disturbance and its mean value (computed along the plumbline) respectively.

Sjöberg (2010) recommended computing the topographic term (i. e. , the second constituent on the right-hand side of Eq. 4. 16) in terms of the gravitational potential generated by a planar terrain roughness term and formulated more rigorously the gravimetric term (i. e. , the third term on the right-hand side of Eq. 4. 16) by means of the vertical gradient of the no-topography gravity anomaly Δg^{NT}. He presented the following expression

$$H^N(\Omega) - H^O(\Omega) \cong \frac{H^O(\Omega)}{\overline{\gamma}(\Omega)}\Delta g^{RPB}(r_t,\Omega) + \frac{V^{TC}(r_g,\Omega) - V^{TC}(r_t,\Omega)}{\overline{\gamma}(\Omega)}$$
$$- \frac{H^2(\Omega)}{2\gamma(H^N,\phi)}\frac{\partial \Delta g^{NT}}{\partial h}\bigg|_{h=H}. \tag{4. 17}$$

Alternatively, the geoid and quasigeoid definitions can be compared in deriving the geoid-to-quasigeoid correction. For this purpose, we define these two quantities by means of applying Bruns' (1878) theorem.

The geoid height N is defined by (e. g. , Heiskanen and Moritz, 1967)

$$N(\Omega) = \frac{T(r_g,\Omega)}{\gamma_0(\phi)}, \tag{4. 18}$$

where the disturbing potential T is stipulated at the geoid surface (r_g,Ω).

Molodensky (1945, 1948) defined the height anomaly ζ in the following form (see also Molodensky et al. , 1960)

$$\zeta(\Omega) = \frac{T(r_t,\Omega)}{\gamma(H^N,\phi)} , \qquad (4.19)$$

where T is the disturbing potential at the topographic surface (r_t,Ω), and the normal gravity γ is evaluated at the telluroid (H^N,ϕ) which is vertically displaced from the ellipsoid surface by the normal height H^N (see Fig. 4.1). Note that the non-linear terms in Eqs. (4.18) and (4.19) are disregarded.

5. Geoid-to-Quasigeoid Correction

Santos et al. (2006) demonstrated that the computation of the geoid-to-quasigeoid correction according to Eq. (4.14) can yield errors up to a few decimetres in the mountainous regions. To reduce these errors, the geoid-to-quasigeoid correction should be computed more accurately using available terrain and density models. The expressions for an accurate determination of the geoid-to-quasigeoid correction are given in the spatial and spectral domains. The computation of this correction can then be practically realized according to the remove-compute-restore numerical scheme, where the long-to-medium wavelength part of this correction is computed based on applying methods for a spherical harmonic analysis and synthesis of the global gravity and crustal structure models, and its residual (higher-frequency) part is determined from observed gravity data and detailed topographic and geological information.

5.1   Relation between geoid and quasigeoid

From Eqs. (4.18) and (4.19), the geoid-to-quasigeoid correction χ is given by

$$\chi(\Omega) = N(\Omega) - \zeta(\Omega) = \frac{T(r_g, \Omega)}{\gamma_0(\phi)} - \frac{T(r_t, \Omega)}{\gamma(H^N, \phi)}. \qquad (5.1)$$

After some algebra, Eq. (5.1) is rewritten as (e. g. , Tenzer et al. , 2006)

$$\chi(\Omega) = \frac{1}{\gamma_0(\phi)}[T(r_g,\Omega) - T(r_t,\Omega)] + \frac{T(r_t,\Omega)}{\gamma(H^N,\phi)}\left[\frac{\gamma(H^N,\phi)}{\gamma_0(\phi)} - 1\right]$$

$$= \frac{1}{\gamma_0(\phi)}[T(r_g,\Omega) - T(r_t,\Omega)] + \zeta(\Omega)\frac{\gamma(H^N,\phi) - \gamma_0(\phi)}{\gamma_0(\phi)}.$$

$$(5.2)$$

The geoid-to-quasigeoid correction in Eq. (5.2) is described by two constituents related to the disturbing potential difference and the normal gravity difference. The former represents a major contribution to χ. The latter is further defined in terms of the normal vertical gravity gradient.

Disregarding the non-linear changes in the vertical normal gravity gradient, we can write

$$\gamma(H^N,\phi) \approx \gamma_0(\phi) + \frac{\partial\gamma}{\partial h}\bigg|_{\gamma=\gamma_0} H^N(\Omega) . \qquad (5.3)$$

The second constituent on the right-hand side of Eq. (5.2) then becomes

$$\zeta(\Omega)\frac{\gamma(H^N,\phi) - \gamma_0(\phi)}{\gamma_0(\phi)} \cong \frac{1}{\gamma_0(\phi)}\frac{\partial\gamma}{\partial h}\bigg|_{\gamma=\gamma_0} H^N(\Omega)\zeta(\Omega) .$$

$$(5.4)$$

Substituting for $\gamma_0 \approx GM/R^2$ and $\partial\gamma/\partial h \approx -2GM/R^3$ in Eq. (5.4), we get

$$\zeta(\Omega)\frac{\gamma(H^N,\phi) - \gamma_0(\phi)}{\gamma_0(\phi)} \approx -\frac{2GM}{R^3}\frac{R^2}{GM}H^N(\Omega)\zeta(\Omega)$$

$$= -\frac{2}{R}H^N(\Omega)\zeta(\Omega) . \qquad (5.5)$$

As seen in Eq. (5.5), this constituent can readily be computed from the known values of ζ and H^N. Setting $H \approx 8.85 \times 10^3$m and $\zeta \approx \pm 100$m, it can reach ± 0.3 m. The errors due to disregarding the non-linear terms in Eq. (5.3) are less than 1 mm. This is evident from: $(2\gamma_0)^{-1}H^2\zeta\partial^2\gamma/\partial h^2 \approx H^2\zeta/3R^2$, where $\partial^2\gamma/\partial h^2 \approx 6GM/R^4$.

The computational procedure is now reduced to evaluate the disturbing potential difference in Eq. (5.2). For this purpose, the disturbing potential T is defined as a function of the gravity disturbance δg, i.e., $\delta g \cong - \partial T/\partial r$. With reference to Tenzer (2004), we can write

$$T(r_g,\Omega) - T(r_t,\Omega) = - \int_{r=r_g}^{r_t} \frac{\partial T}{\partial r} dr \cong \int_{r=r_g}^{r_t} \delta g(r,\Omega) dr. \quad (5.6)$$

The integral mean of the gravity disturbance reads

$$\overline{\delta g}(\Omega) = \frac{1}{H^0(\Omega)} \int_{r=r_g}^{r_t} \delta g(r,\Omega) dr. \quad (5.7)$$

From Eqs. (5.6) and (5.7), we get

$$T(r_g,\Omega) - T(r_t,\Omega) = H^0(\Omega) \overline{\delta g}(\Omega). \quad (5.8)$$

Substituting from Eqs. (5.5) and (5.8), the geoid-to-quasigeoid correction in Eq. (5.2) becomes

$$X(\Omega) = \frac{1}{\gamma_0(\phi)} H^0(\Omega) \overline{\delta g}(\Omega) - \frac{2}{R} H^N(\Omega) \zeta(\Omega). \quad (5.9)$$

The orthometric height in Eq. (5.9) can be replaced by the normal height, and vice versa, without any significant influence on the accuracy; for $H^0 - H^N = \pm 5 \mathrm{m}$ and $\overline{\delta g} = 500$ mGal, the error of X is only ~ 2.5 mm.

5.2 Computation in spatial domain

The disturbing potential difference in Eq. (5.8) is evaluated from the observed gravity disturbances/anomalies by applying four numerical steps consisting of: (i) the forward modelling of the topographic gravity correction, (ii) the inverse solution to discretized Green's integral equations, (iii) the solution to discretized Poisson's integral equation, and (iv) the forward modelling of the topographic potential difference. These

22

numerical steps are summarized below.

By analogy with the procedure described in Tenzer et al. (2005), we separate the disturbing potential T into the components which can be calculated separately from gravity data and available topographic and atmospheric density models. We then write

$$T(r,\Omega) = V^T(r,\Omega) + V^A(r,\Omega) + T^{NT}(r,\Omega) , \qquad (5.10)$$

where V^T and V^A are the topographic and atmospheric potentials respectively, and T^{NT} is the no-topography disturbing potential (Vaníček et al. , 2005). Moreover, it is convenient to separate the topographic potential into the components V^{T,ρ^T} and $V^{T,\delta\rho^T}$ which are related to the constant and anomalous topographic density distribution, i. e. , $\rho^T = const.$ and $\delta\rho^T = \rho^T(r_1\Omega) - \rho^T$, where ρ^T is the actual topographic density. Hence

$$V^T(r,\Omega) = V^{T,\rho^T}(r,\Omega) + V^{T,\delta\rho^T}(r,\Omega) . \qquad (5.11)$$

Substituting from Eqs. (5.10) and (5.11) to Eq. (5.8), we arrive at

$$
\begin{aligned}
T(r_g,\Omega) &- T(r_t,\Omega) \\
&\cong [V^{T,\rho^T}(r_g,\Omega) - V^{T,\rho^T}(r_t,\Omega)] + [V^{T,\delta\rho^T}(r_g,\Omega) - V^{T,\delta\rho^T}(r_t,\Omega)] \\
&+ [V^A(r_g,\Omega) - V^A(r_t,\Omega)] + [T^{NT}(r_g,\Omega) - T^{NT}(r_t,\Omega)] .
\end{aligned}
$$

$$(5.12)$$

Whereas the atmospheric and topographic potentials V^A , V^{T,ρ^T} and $V^{T,\delta\rho^T}$ in Eq. (5.12) can directly be computed from the topographic data and the atmospheric and topographic density models by applying the gravimetric forward modelling, the evaluation of the no-topography disturbing potential difference from the observed gravity data requires the application of the atmospheric and topographic gravity corrections followed by the gravity-to-potential conversion of the non-topographic part of the gravity field with the additional upward continuation.

The no-topography gravity disturbances δg^{NT} at the topographic surface (or above) are obtained from the observed gravity disturbances δg after applying the direct topographic and atmospheric effects, i. e. ,

$$\delta g^{NT}(r_t,\Omega) = \delta g(r_t,\Omega) - g^{T,\rho^T}(r_t,\Omega) - g^{T,\delta\rho^T}(r_t,\Omega) - g^A(r_t,\Omega) ,$$

(5. 13)

where g^A is the atmospheric attraction, and the topographic attraction g^T consists of two components g^{T,ρ^T} and $g^{T,\delta\rho^T}$ which are evaluated again for the reference and anomalous topographic density distribution $\rho^T = const.$ and $\delta\rho^T$ respectively.

The computation of the no-topography gravity anomalies Δg^{NT} from the observed gravity anomalies Δg is realized by applying the direct and secondary indirect effects of topography and atmosphere (cf. Vaníček et al. , 2005)

$$\Delta g^{NT}(r_t,\Omega) = \Delta g(r_t,\Omega) - g^{T,\rho^T}(r_t,\Omega) - g^{T,\delta\rho^T}(r_t,\Omega)$$

$$- g^A(r_t,\Omega) + \frac{2}{r_t(\Omega)}V^{T,\rho^T}(r_t,\Omega) + \frac{2}{r_t(\Omega)}V^{T,\delta\rho^T}(r_t,\Omega)$$

$$+ \frac{2}{r_t(\Omega)}V^A(r_t,\Omega) .$$

(5. 14)

As seen in Eqs. (5. 13) and (5. 14), the direct and secondary indirect topographic effects are defined as g^T and $2r^{-1}V^T$ respectively. Equivalently, the values g^A and $2r^{-1}V^A$ represent the direct and secondary indirect atmospheric effects respectively.

5. 2. 1 Topographic component

The topographic potential V^{T,ρ^T} is evaluated at a point (r,Ω) by means of solving Newton's volumetric integral (e. g. , Martinec, 1998)

$$V^{T,\rho^T}(r,\Omega) \cong G\rho^T \iint_\Phi \int_{r'=R}^{R+H'} \ell^{-1}(r,\psi,r')r'^2 dr'd\Omega' ,$$

(5. 15)

where ℓ is the Euclidean spatial distance of two points (r,Ω) and (r',Ω'), ψ is the respective spherical distance, $d\Omega' = \cos\phi'd\phi'd\lambda'$ is the infinitesimal surface element, and $\Phi = \{\ \Omega' = (\phi',\lambda'):\phi' \in [-\pi/2,\pi/2] \wedge \lambda' \in [0,2\pi)\ \}$ is the full spatial angle. The reference topographic density of $\rho^T = 2,670$ kg/m^3 is typically attributed to the average density of the upper continental crust (cf. Hinze, 2003).

By analogy with Eq. (5.15), we define the topographic potential $V^{T,\delta\rho^T}$ generated by the anomalous topographic density $\delta\rho^T$ as follows (e.g., Martinec, 1998)

$$V^{T,\delta\rho^T}(r,\Omega) \cong G\iint_{\Phi} \int_{r'=R}^{R+H'} \delta\rho^T(r',\Omega')\ell^{-1}(r,\psi,r')r'^2dr'd\Omega'$$

(5.16)

The topographic attractions g^{T,ρ^T} and $g^{T,\delta\rho^T}$ (defined approximately as a negative radial derivative of the respective topographic potentials V^{T,ρ^T} and $V^{T,\delta\rho^T}$) are given by (ibid.)

$$g^{T,\rho^T}(r,\Omega) \cong -\frac{\partial V^{T,\rho^T}}{\partial r} \cong -G\rho^T\iint_{\Phi}\int_{r'=R}^{R+H'}\frac{\partial\ell^{-1}(r,\psi,r')}{\partial r}r'^2dr'd\Omega'\ ,$$

(5.17)

and

$$g^{T,\delta\rho^T}(r,\Omega) \cong -\frac{\partial V^{T,\delta\rho^T}}{\partial r}$$

$$\cong -G\iint_{\Phi}\int_{r'=R}^{R+H'}\delta\rho^T(r',\Omega')\frac{\partial\ell^{-1}(r,\psi,r')}{\partial r}r'^2dr'd\Omega'\ .$$

(5.18)

In Eqs. (5.15-5.18) and all equations hereafter, we apply the spherical approximation. The relative errors of $\sim 0.3\%$ are to be expected in the computed gravity field quantities due to disregarding the Earth's flattening (cf. Heiskanen and Moritz, 1967). Moreover, the strict definition of the height system (i.e., either by the orthometric or normal heights) is not required in

25

computing the geoid-to-quasigeoid correction due to the fact that topographic information is retrieved from digital terrain models (DTMs). Hence, we refer here to the topographic height H instead of specifically to H^N or H^O. We then define the geocentric radii of the topographic and geoid surface as $r_t \cong R + H$ and $r_g \cong R$ respectively, and the integral limit of the radial integration as $R \leqslant r' \leqslant R + H'$.

5. 2. 2　Atmospheric component

Considering only the radially distributed atmospheric mass density $\rho^A(r)$, the atmospheric potential V^A is defined as (Sjöberg, 1999, 2001; Novák, 2000)

$$V^A(r,\Omega) \cong G \iint_{\Phi} \int_{r'=R+H'}^{R+H_{\max}} \rho^A(r') \ell^{-1}(r,\psi,r') r'^2 dr' d\Omega' +$$

$$\iint_{\Phi} \int_{r'=R+H_{\max}}^{r_{\lim}} \rho^A(r') \ell^{-1}(r,\psi,r') r'^2 dr' d\Omega' . \qquad (5.19)$$

The volumetric integration domain within the Earth's atmosphere in Eq. (5. 19) is divided into the atmospheric spherical shell and the atmospheric roughness term. The atmospheric spherical shell is defined between the upper limit of topography $H_{\max}$ (chosen as a maximum topographic height) and the upper limit of atmosphere $r_{\lim}$ (where the gravitational effect of atmospheric density becomes negligible, typically ~ 50 km above sea level). The atmospheric roughness term is enclosed by the topographic surface and the upper limit of topography.

For $r < R + H_{\max}$, the gravitational potential of atmospheric spherical shell (given by the second integral on the right-hand side of Eq. 5. 19) is constant (e. g. , MacMillan, 1930)

$$G \iint_{\Phi} \int_{r'=R+H_{\max}}^{r_{\lim}} \rho^A(r') \ell^{-1}(r,\psi,r') r'^2 dr' d\Omega'$$

$$= 4\pi G \int_{r'=R+H_{\max}}^{r_{\lim}} \rho^A(r') r' dr' , \qquad (5.20)$$

The atmospheric potential difference in Eq. (5. 12) then reduces to

$$V^A(r_g,\Omega) - V^A(r_t,\Omega) \cong G\iint_{\Phi} \int_{r'=R+H'}^{R+H_{max}} \rho^A(r') \left[\ell^{-1}(R,\psi,r') \right.$$
$$\left. - \ell^{-1}(R+H,\psi,r') \right] r'^2 dr' d\Omega' . \quad (5.21)$$

The computation of the atmospheric potential difference in Eq. (5. 21) can be realized using the standard model of the static atmosphere (ISO 2533: 1975) . Tenzer et al. (2005) demonstrated, however, that the atmospheric effect on the geoid-to-quasigeoid correction is less than 1 mm and thus completely negligible.

5. 2. 3 Non-topographic component

The computation of the non-topographic part of the geoid-to-quasigeoid correction (i. e. , the effect of mass density heterogeneities distributed below the geoid surface) is realized in two steps. First, the harmonic downward continuation is applied by means of solving the inverse to discretized Green's integral equations. In this numerical step, the values of T^{NT} are determined on the geoid surface from the values δg^{NT} or Δg^{NT} at the topographic surface. The Green integrals read (Tenzer et al. , 2006)

$$r\delta g^{NT}(r,\Omega) = -\frac{r}{4\pi}\iint_{\Phi} \frac{\partial P(r,\psi,R)}{\partial r} T^{NT}(r'_g,\Omega') d\Omega' (r \geqslant R) ,$$

$$(5.22)$$

and

$$r\Delta g^{NT}(r,\Omega) = -\frac{r}{4\pi}\iint_{\Phi}\left[\frac{\partial P(r,\psi,R)}{\partial r} + \frac{2}{r}P(r,\psi,R)\right]$$
$$T^{NT}(r'_g,\Omega') d\Omega' \quad (r \geqslant R) , \quad (5.23)$$

where P is the Poisson kernel. We note here that the products $r\,\delta g^{NT}$ and $r\,\Delta g^{NT}$ are harmonic above the geoid surface, i. e. , they satisfy Laplace equation: $\forall r > R : \Delta(r \delta g^{NT}) = 0$ and $\forall r > R :$ $\Delta(r \Delta g^{NT}) = 0$. The harmonic upward continuation, realized by solving discretized Poisson's integral equation, is then applied to evaluate the no-topography disturbing potential T^{NT} at the topographic surface (Tenzer et al., 2006). The Poisson integral reads

$$T^{NT}(r,\Omega) = \frac{1}{4\pi} \iint_{\Phi} P(r,\psi,R) T^{NT}(r'_g,\Omega') d\Omega' \quad (r \geqslant R).$$

$$(5.24)$$

For $r \geqslant R$, the Poisson kernel P and the respective Green kernels $\partial P/\partial r$ and $\partial P/\partial r + 2P/r$ in Eqs. (5.22-5.24) are given by

$$P(r,\psi,R) = R \frac{r^2 - R^2}{\ell^3(r,\psi,R)}, \quad (5.25)$$

$$\frac{\partial P(r,\psi,R)}{\partial r} = \frac{2rR}{\ell^3(r,\psi,R)} - 3R \frac{r^2 - R^2}{\ell^5(r,\psi,R)}(r - R\cos\psi),$$

$$(5.26)$$

$$\frac{\partial P(r,\psi,R)}{\partial r} + \frac{2}{r} P(r,\psi,R)$$

$$= \frac{2R}{r} \frac{r^2 - R^2}{\ell^3(r,\psi,R)} - 3R \frac{r^2 - R^2}{\ell^5(r,\psi,R)}(r - R\cos\psi). \quad (5.27)$$

An alternative approach to compute the no-topography disturbing potential difference was proposed by Tenzer et al. (2005). They applied the downward continuation of the no-topography gravity disturbances based on solving the inverse to discretized Poisson's integral equations (i. e., the inverse solution to Dirichlet's boundary-value problem; Martinec, 1996). The resulting no-topography gravity disturbances at the geoid are then used to evaluate the radial integral of discretized Poisson's integral equation. The Poisson integral for δg^{NT} reads

$$r \delta g^{NT}(r,\Omega) = \frac{R}{4\pi} \iint_{\Phi} P(r,\psi,R) \delta g^{NT}(r'_g,\Omega') d\Omega' \quad (r \geqslant R).$$

$$(5.28)$$

The radial integral of Poisson's integral is given by (Tenzer et al. , 2005)

$$T^{NT}(r_g,\Omega) - T^{NT}(r_t,\Omega) \cong \int_{r=R}^{R+H} \delta g^{NT}(r,\Omega)dr$$

$$= \frac{R}{4\pi}\iint_{\Phi}\int_{r=R}^{R+H} r^{-1}P(r,\psi,R)dr\delta g^{NT}(r'_g,\Omega')d\Omega' \quad (r \geq R) , \quad (5.29)$$

where

$$\int_r r^{-1}P(r,\psi,R)dr = -2R\ell^{-1}(r,\psi,R) + \ln\left|\frac{R - r\cos\psi + \ell(r,\psi,R)}{r\sin\psi}\right| .$$

$$(5.30)$$

Since gravity anomalies are still the most commonly used gravity data type, it is convenient to compute the geoid-to-quasigeoid correction from observed gravity anomalies without their additional conversion to gravity disturbances. Such procedure was discussed in Tenzer et al. (2006) and Sjöberg (2006). According to their approach, the no-topography gravity anomalies are first downward continued onto the geoid. The generalized Stokes problem is then solved to evaluate the respective disturbing potential values at the topographic surface and geoid in order to obtain the (no-topography) disturbing potential difference. By analogy with Eq. (5. 28), the Poisson integral for Δg^{NT} reads

$$r\Delta g^{NT}(r,\Omega) = \frac{R}{4\pi}\iint_{\Phi}P(r,\psi,R)\Delta g^{NT}(r'_g,\Omega')d\Omega' \quad (r \geq R) .$$

$$(5.31)$$

The solution to generalized Stokes' problem for finding the disturbing potential difference is defined as follows

$$T^{NT}(r_g,\Omega) - T^{NT}(r_t,\Omega)$$

$$\cong \frac{R}{4\pi}\iint_{\Phi}[S(\psi) - S(r_t,\psi)]\Delta g^{NT}(r'_g,\Omega')d\Omega' , \quad (5.32)$$

where the closed analytical formulas of the Stokes kernels $S(\psi)$ and $S(r,\psi)$ are given by (cf. Heiskanen and Moritz, 1967)

$$S(\psi) = \ \operatorname{cosec} \frac{\psi}{2} - 6\sin\frac{\psi}{2} + 1 - 5\cos\psi - 3\cos\psi\ln\left(\sin\frac{\psi}{2} + \sin^2\frac{\psi}{2}\right) ,$$

$$(5.33)$$

and

$$S(r,\psi) = \frac{2R}{\ell(r,\psi,R)} + \frac{R}{r} - 3\,\frac{R}{r^2}\ell(r,\psi,R)$$
$$- \left(\frac{R}{r}\right)^2 \cos\psi\left[5 + 3\ln\frac{r - R\cos\psi + \ell(r,\psi,R)}{2r}\right].$$

$$(5.34)$$

Whereas the gravity-to-potential conversion of the non-topographic part of the gravity field is realized through the downward continuation procedure of solving the inverse to discretized Green's integral equations (Eqs. 5.22 and 5.23), in the alternative approaches this conversion is realized after the gravity downward continuation by solving either the radial integral of discretized Poisson's integral equation (for the values of δg^{NT}; Eq. 5.29) or the generalized Stokes problem (for the values of Δg^{NT}; Eq. 5.32). Tenzer and Novák (2008) demonstrated that the downward continuation procedure by means of solving the inverse to discretized Green's integral equations (Eqs. 5.22 and 5.23) is numerically more stable than Poisson's downward continuation procedure (Eqs. 5.28 and 5.31).

In the final step, the disturbing potential difference is obtained from the respective non-topographic value by adding the topographic potential difference (see Eq. 5.12). The computation of the topographic potential difference can be realized by solving either Newton's integral for the potential values or the radial integral of Newton's integral for the topographic attraction.

5. 3 Computation in spectral domain

The computation of the geoid-to-quasigeoid correction in the spectral domain is realized based on utilizing methods for a spherical analysis and synthesis of gravity field and topographic density structure. The spectral expressions can be applied to evaluate the long-to-medium wavelength part of this correction, while the residual (higher-frequency) contribution could be computed in the spatial domain from local gravity data and detailed topographic and crustal density models (according to the expressions summarized in Section 5. 2).

5. 3. 1 Topographic potential difference (of reference density)

In the spectral domain, the topographic potential difference (i. e. , the first constituent on the right-hand side of Eq. 5. 12) can be defined in terms of Newton's volumetric integral for the external and internal convergence domains. Alternatively, the topographic potential difference can be described by means of the analytical downward continuation of the topographic potential at the topographic surface and applying the topographic bias (cf. Årgen, 2004; Sjöberg, 2007; Vermeer, 2008). The topographic potential difference then becomes

$$V^{T,\rho^T}(r_g,\Omega) - V^{T,\rho^T}(r_t,\Omega) = \lim_{r \to R^+} V_e^{T,\rho^T}(r,\Omega) - V_e^{T,\rho^T}(r_t,\Omega)$$
$$+ \lim_{r \to R^-} V_i^{T,\rho^T}(r,\Omega) - \lim_{r \to R^+} V_e^{T,\rho^T}(r,\Omega) .$$

$$(5. 35)$$

The(negative) analytical continuation term $\lim_{r \to R^+} V_e^{T,\rho^T}(r,\Omega) - V_e^{T,\rho^T}(r_t,\Omega)$ is defined by means of Newton's volumetric integral V_e^{T,ρ^T} for the external convergence domain (e. g. , Hobson, 1931;

Kellogg, 1929; Moritz, 1980) as follows

$$\lim_{r \to R^+} V_e^{T,\rho^T}(r,\Omega) - V_e^{T,\rho^T}(r_t,\Omega)$$

$$= \frac{GM}{R} \sum_{n=0}^{\bar{n}} \left[1 - \left(1 + \frac{H}{R} \right)^{-n-1} \right] \sum_{m=-n}^{n} V_{n,m}^{T,\rho^T} Y_{n,m}(\Omega) , \qquad (5.36)$$

where $\bar{n}$ is a maximum degree of series expansion. The potential coefficients $V_{n,m}^{T,\rho^T}$ in Eq. (5.36) read

$$V_{n,m}^{T,\rho^T} = \frac{3}{2n+1} \frac{\rho^T}{\bar{\rho}^{\,\text{Earth}}} \sum_{k=0}^{n+2} \binom{n+2}{k} \frac{1}{k+1} \frac{H_{n,m}^{(k+1)}}{R^{k+1}} , \qquad (5.37)$$

where $\bar{\rho}^{\,\text{Earth}} = 5,500 \text{ kg m}^{-3}$ is the Earth's mean mass density (e. g. , Novák, 2010). The analytical downward continuation of V_e^{T,ρ^T} is permissible due to assuming a constant topographic density ρ^T. It is worth mentioning that the analytical continuation can also be applied for a lateral density distribution.

The Laplace harmonics H_n of the topography are defined by the following integral convolution

$$H(\Omega) = \sum_{n=0}^{n} H_n(\Omega) , \quad H_n(\Omega) = \frac{2n+1}{4\pi} \iint_{\Phi} H' P_n(t) d\Omega'$$

$$= \sum_{m=-n}^{n} H_{n,m} Y_{n,m}(\Omega) , \qquad (5.38)$$

where $H_{n,m}$ are the topographic coefficients. The corresponding higher-order harmonics $\{ H_n^{(k)} : k = 2,3,\cdots \}$ read

$$H_n^{(k)}(\Omega) = \frac{2n+1}{4\pi} \iint_{\Phi} H'^k P_n(t) d\Omega'$$

$$= \sum_{m=-n}^{n} H_{n,m}^{(k)} Y_{n,m}(\Omega) . \qquad (5.39)$$

Sjöberg (2007) defined the topographic bias as the difference between the downward-continued gravitational potential $\lim_{r \to R^+} V_e^{T,\rho^T}(r,\Omega)$ and the topographic potential computed for the internal convergence domain $\lim_{r \to R^-} V_i^{T,\rho^T}(r,\Omega)$. The (negative) topographic bias in Eq. (5.35) is then given by

32

$$\lim_{r \to R^-} V_i^{T,\rho^T}(r,\Omega) - \lim_{r \to R^+} V_e^{T,\rho^T}(r,\Omega) = -\frac{GM}{R} \sum_{n=0}^{\bar{n}} \sum_{m=-n}^{n} V_{n,m}^{\text{bias}} Y_{n,m}(\Omega)$$

$$(5.40)$$

where the topographic bias coefficients $V_{n,m}^{\text{bias}}$ read

$$V_{n,m}^{\text{bias}} = 3\frac{\rho^T}{\rho^{\text{Earth}}} \sum_{k=1}^{2} \frac{1}{k+1} \frac{H_{n,m}^{(k+1)}}{R^{k+1}}.$$

$$(5.41)$$

The topographic bias represents the discontinuity of the gravitational potential generated by the spherical Bouguer shell at its lower topographic bound $r = R$. The gravitational potential of the spherical roughness term is zero, because its values evaluated for $r \to R^+$ and $r \to R^-$ are the same (cf. Sjöberg, 2007).

5. 3. 2 Topographic potential difference (of anomalous density)

The topographic potential $V^{T,\delta\rho^T}$ generated by the anomalous topographic density distribution can be defined in terms of the Laplace harmonics of $\delta\rho^T H_n$ and their higher-order terms $\{\delta\rho^T H_n^{(k)} : k = 2, 3, \cdots\}$, if the density function $\delta\rho^T$ describes uniquely the anomalous density distribution within the whole topography. This is possible only if the radial density distribution is continuous everywhere within the topography. In reality, however, the density distribution within the upper continental crust comprises density interfaces, such as below the polar ice sheets or sedimentary basins and the underlying bedrock. For this reason, we separate the anomalous topographic density into particular continental crustal components (above sea level) attributed to the continental water bodies (lakes), ice, sediments and remaining anomalous topographic masses. This description utilizes an arbitrary volumetric mass layer (within the topography) with a variable depth and thickness. The upper and lower bounds of this layer are defined by the parameters H_U and H_L

33

of the vertical displacement relative to the geoid surface, which is approximated by the radius R. In the most general case, the actual density ρ within every volumetric mass layer can be approximated by the laterally distributed radial density variation model using the following polynomial function (Tenzer et al., 2012b)

$$\rho(r,\Omega) = \rho(H_U,\Omega) + \beta(\Omega) \sum_{i=1}^{I} \alpha_i(\Omega) \ (r - R)^i,$$
$$for \quad R + H_U(\Omega) \geqslant r > R + H_L(\Omega), \quad\quad (5.42)$$

where $\rho(H_U,\Omega)$ is the (nominal) lateral density at the upper bound H_U and location Ω. The radial density change at this location Ω with respect to the nominal density $\rho(H_U,\Omega)$ is described by the parameters β and $\{\alpha_i : i = 1,2,\cdots,I\}$, where I is the maximum order of the radial density distribution function.

Since the anomalous topographic density $\delta\rho^T$ is defined with respect to the reference topographic density ρ^T, we define the density contrast within a volumetric mass layer by

$$\delta\rho(r,\Omega) = \rho(r,\Omega) - \rho^T$$
$$= \delta\rho(H_U,\Omega) + \beta(\Omega) \sum_{i=1}^{I} \alpha_i(\Omega) \ (r - R)^i,$$
$$for \quad R + H_U(\Omega) \geqslant r > R + H_L(\Omega), \quad\quad (5.43)$$

where $\delta\rho(H_U,\Omega)$ is the nominal value of the lateral density contrast at a location (H_U,Ω).

Most of the Earth's inner density structures can be approximated with a sufficient accuracy by the constant, laterally or radially varying density distribution models. The uniform density contrast model is, for example, suitable to approximate the density contrast of continental ice sheets and continental water bodies. The lateral density model (of multiple layers) can be implemented to describe the density contrast structures within the sediments and remaining continental crust. Alternatively, the increasing density with depth within sedimentary basins due to compaction can be described more realistically by assuming a

34

depth-dependent density model (e. g. , Artemjev et al. , 1994).

The gravitational potential of mass density contrast layer is defined for the external convergence domain in the following form (cf. Tenzer et al. , 2012b, 2012c, 2012d)

$$V_e^{\delta\rho}(r,\Omega) = \frac{GM}{R} \sum_{n=0}^{\bar{n}} \left(\frac{R}{r}\right)^{n+1} \sum_{m=-n}^{n} {}_eV_{n,m}^{\delta\rho} Y_{n,m}(\Omega) \ , \qquad (5.44)$$

where the potential coefficients ${}_eV_{n,m}^{\delta\rho}$ read

$$_eV_{n,m}^{\delta\rho} = \frac{3}{2n+1} \frac{1}{\bar{\rho}^{\text{Earth}}} \sum_{i=0}^{I} ({}_eFu_{n,m}^{(i)} - {}_eFl_{n,m}^{(i)}) \ . \qquad (5.45)$$

The numerical coefficients $\{ \ {}_eFl_{n,m}^{(i)}, {}_eFu_{n,m}^{(i)} : i = 0,1,\cdots,I \}$ in Eq. (5.45) are given by

$$_eFl_{n,m}^{(i)} = \sum_{k=0}^{n+2} \binom{n+2}{k} \frac{1}{k+1+i} \frac{L_{n,m}^{(k+1+i)}}{R^{k+1}} \qquad (i = 0,1,\cdots,I) \ ,$$

$$(5.46)$$

and

$$_eFu_{n,m}^{(i)} = \sum_{k=0}^{n+2} \binom{n+2}{k} \frac{1}{k+1+i} \frac{U_{n,m}^{(k+1+i)}}{R^{k+1}} \qquad (i = 0,1,\cdots,I) \ .$$

$$(5.47)$$

The spherical lower-bound and upper-bound functions L_n and U_n of degree n and their higher-order terms read

$$L_n^{(k+1+i)}(\Omega) = \begin{cases} \dfrac{2n+1}{4\pi} \iint\limits_{\Phi} \delta\rho(H'_U,\Omega') H_L^{k+1}(\Omega') P_n(t) d\Omega' \\[2mm] = \displaystyle\sum_{m=-n}^{n} L_{n,m}^{(k+1)} Y_{n,m}(\Omega) \qquad\qquad i = 0 \\[5mm] \dfrac{2n+1}{4\pi} \iint\limits_{\Phi} \beta(\Omega') \alpha_i(\Omega') H_L^{k+1+i}(\Omega') P_n(t) d\Omega' \\[2mm] = \displaystyle\sum_{m=-n}^{n} L_{n,m}^{(k+1+i)} Y_{n,m}(\Omega) \qquad i = 1,2,\cdots,I \end{cases} \qquad (5.48)$$

and

$$U_n^{(k+1+i)}(\Omega) = \begin{cases} \dfrac{2n+1}{4\pi} \iint\limits_{\Phi} \delta\rho(H'_U,\Omega')H_U^{k+1}(\Omega')P_n(t)\,d\Omega' \\[2mm] = \displaystyle\sum_{m=-n}^{n} U_{n,m}^{(k+1)} Y_{n,m}(\Omega) \qquad i=0 \\[4mm] \dfrac{2n+1}{4\pi} \iint\limits_{\Phi} \beta(\Omega')\alpha_i(\Omega')H_U^{k+1+i}(\Omega')P_n(t)\,d\Omega' \\[2mm] = \displaystyle\sum_{m=-n}^{n} U_{n,m}^{(k+1+i)} Y_{n,m}(\Omega) \qquad i=1,2,\cdots,I \end{cases} \tag{5.49}$$

By analogy with Eq. (5.44), the gravitational potential of mass density contrast layer is defined for the internal convergence domain in the following form

$$\lim_{r\to R^-} V_i^{\delta\rho}(r,\Omega) = \frac{GM}{R}\sum_{n=0}^{\bar{n}}\sum_{m=-n}^{n} {}_i V_{n,m}^{\delta\rho} Y_{n,m}(\Omega), \tag{5.50}$$

where the potential coefficients ${}_i V_{n,m}^{\delta\rho}$ are given by

$${}_i V_{n,m}^{\delta\rho} = \frac{3}{2n+1}\frac{1}{\rho^{\text{Earth}}}\sum_{i=0}^{I}({}_i Fu_{n,m}^{(i)} - {}_i Fl_{n,m}^{(i)}). \tag{5.51}$$

The coefficients $\{{}_i Fl_{n,m}^{(i)}, {}_i Fu_{n,m}^{(i)} : i = 0,1,\cdots,I\}$ in Eq. (5.51) read

$${}_i Fl_{n,m}^{(i)} = \sum_{k=0}^{\infty}(-1)^k\binom{n+k-2}{k}\frac{1}{k+1+i}\frac{L_{n,m}^{(k+1+i)}}{R^{k+1}} \quad (i=0,1,\cdots,I), \tag{5.52}$$

and

$${}_i Fu_{n,m}^{(i)} = \sum_{k=0}^{\infty}(-1)^k\binom{n+k-2}{k}\frac{1}{k+1+i}\frac{U_{n,m}^{(k+1+i)}}{R^{k+1}} \quad (i=0,1,\cdots,I). \tag{5.53}$$

With reference to Eqs. (5.44 and 5.50), the topographic potential difference of a volumetric mass density contrast layer is obtained in the following form

$$\lim_{r\to R^-} V_i^{\delta\rho}(r,\Omega) - V_e^{\delta\rho}(r_t,\Omega)$$

$$= \frac{GM}{R} \sum_{n=0}^{\bar{n}} \sum_{m=-n}^{n} \left[{}_i V_{n,m}^{\delta\rho} - \left(\frac{R}{R+H} \right)^{n+1} {}_e V_{n,m}^{\delta\rho} \right] Y_{n,m}(\Omega) . \qquad (5.54)$$

As seen in Eq. (5.54), the potential difference of a volumetric mass density contrast layer is defined using the expressions derived for the external and internal convergence domains, while the expression for computing the topographic potential difference (of the reference density) in Eq. (5.35), cf. also Eqs. (5.36) and (5.40), combines the analytical continuation term and the topographic bias.

5. 3. 3   No-topography disturbing potential difference

To evaluate the non-topographic part of the Earth's gravity field, the topographic potential V^T is first subtracted from the disturbing potential T , both referred at the topographic surface (or above). With reference to Eq. (3.5), the disturbing potential T at the topographic surface (r_t, Ω) reads

$$T(r_t, \Omega) = \frac{GM}{R} \sum_{n=0}^{\bar{n}} \sum_{m=-n}^{n} \left(\frac{R}{r_t} \right)^{n+1} T_{n,m} Y_{n,m}(\Omega) . \qquad (5.55)$$

Taking into consideration the decomposition of the topographic potential in Eq. (5.11), the no-topography disturbing potential T^{NT} at the topographic surface (r_t, Ω) is defined as

$$T^{NT}(r_t, \Omega) = T(r_t, \Omega) - V_e^{T,\rho^T}(r_t, \Omega) - \sum_j V_{e_j^{\delta\rho}}(r_t, \Omega) ,$$

$$(5.56)$$

where j is the summation index of the volumetric mass density contrast layers applied to describe the anomalous density distribution within the topography. From Eq. (5.56), we have

$$T^{NT}(r_t, \Omega) = \frac{GM}{R} \sum_{n=0}^{\bar{n}} \sum_{m=-n}^{n} \left(\frac{R}{r_t} \right)^{n+1} (T_{n,m} - V_{n,m}^{T,\rho^T} + \sum_j {}_e V_{n,mj}^{\delta\rho}) Y_{n,m}(\Omega) .$$

$$(5.57)$$

The computation of the no-topography gravity field quantities according to Eq. (5. 57) is realized by generating Stokes' coefficients of the no-topography disturbing potential, i. e. , $T_{n,m}^{NT} =$ $T_{n,m} - V_{n,m}^{T,\rho^{T}} + \sum_{j} {}_{e}V_{n,mj}^{\delta\rho}$. These coefficients are then used to compute the no-topography disturbing potential difference from

$$T^{NT}(r_{g},\Omega) - T^{NT}(r_{t},\Omega) = \frac{GM}{R} \sum_{n=0}^{\bar{n}} \sum_{m=-n}^{n} \left[1 - \left(1 + \frac{H}{R} \right)^{-n-1} \right] T_{n,m}^{NT} Y_{n,m}(\Omega) .$$

$$(5.58)$$

As seen in Eqs. (5. 55-5. 58) , the computation of the non-topographic part of the gravity field in the spectral domain is realized by means of the potential field quantities. In the spatial domain, on the other hand, the observed gravity data have to be converted into the potential values.

5. 4   Discussion

The generic expression of the geoid-to-quasigeoid correction in Eq. (5. 2) was found based on comparing the definitions of the geoidal height (Eq. 4. 18) and the height anomaly (Eq. 4. 19) . This generic expression comprises two constituents related to the disturbing potential difference and the normal gravity difference. It was shown (Eq. 5. 9) that the normal gravity difference can readily be computed (to a high accuracy) from topographic and geoid/quasigeoid information. The computation of the disturbing potential difference is, on the other hand, numerically more complex. It requires the evaluation of the topographic and non-topographic parts of the gravity field individually by applying several numerical steps. Moreover, these numerical steps are, in principle, different when performing the computation in the spatial and spectral domains. In the spatial domain, the observed gravity data are first corrected for the gravitational effect of topography. The corrected gravity data (i. e. , the no-topography gravity disturbances/anomalies) are then used to compute the no-topography disturbing potential at the geoid surface by solving the inverse to discretized Green's integral equations. These values are then used to compute the respective potential at the topographic surface in order to obtain the no-topography disturbing potential difference while the topographic potential difference is evaluated by solving Newton's volumetric integral. In the spectral domain, the conversion of gravity to potential is not required. Actually, all computations are realized in potential terms. This obviously significantly simplifies the computation of the non-topographic part of the gravity field in the spectral domain. As seen from Eq. (5.58), the no-topography disturbing potential difference is evaluated from the coefficients of disturbing potential after subtracting the topographic coefficients. These topographic coefficients were defined individually for the constant and variable topographic density distributions. The expressions for computing the topographic potential difference of the reference density (Eq. 5.35) combines the analytical continuation term (Eq. 5.36) and the topographic bias (Eq. 5.40). The potential difference of the volumetric mass density contrast layers, which describe the anomalous density distribution within the topography, was derived by means of the potential coefficients for the external and internal convergence domains (Eq. 5.54).

An accurate computation of the geoid-to-quasigeoid correction according to Eq. (5.9) requires the knowledge of the crustal density structure (within the topography) and topographic information. Consequently, the numerical realization becomes increasingly complex, involving the solutions to Newton's volumetric integrals, and eventually also the downward continuation procedure if the computation is realized in the spatial domain. In contrast, the geoid-to-quasigeoid correction can be computed approximately, according to Eq. (4. 14), only as a function of the simple planar Bouguer gravity anomaly and the topographic height. As already emphasized above, for practical reasons this simple computation would preferably be used for most of the applications. Moreover, this simple definition provides a relatively good accuracy over flat regions. In the mountainous, polar and geologically complex regions, however, the errors in computed values of the geoid-to-quasigeoid correction can reach several decimetres. In this case, the accuracy can be improved significantly by using a more rigorous definition of this correction which incorporates the topographic and density information.

6. Vertical Datum Unification in New Zealand

The combined approach was applied for the experimental unification of the local vertical datums (LVDs) in New Zealand using a relatively sparsely and irregularly distributed GPS, gravity and levelling data. This approach utilised the joint levelling network adjustment and the geopotential-value approach. The levelling data and normal gravity values were used for a joint adjustment of levelling networks at the South and North Islands of New Zealand while fixing the heights of tide gauges in Dunedin and Wellington. Since the geopotential-value approach is based on Molodensky's theory, the newly-adjusted normal-orthometric heights were converted to the normal heights. This conversion was based on applying the cumulative normal to normal-orthometric height correction computed from the spirit levelling and gravity anomaly data. In the absence of the observed gravity data, the gravity anomalies along levelling lines were generated from the global geopotential model EGM2008 (Pavlis et al. 2008). The GPS-levelling data and EGM2008 were then used to estimate the average offsets of the jointly adjusted levelling networks at the North and South Islands with respect to WHS. Since the newly-adjusted levelling networks on both islands still comprised large systematic discrepancies mainly due to the EGM2008 commission and omission errors and systematic errors in levelling data, a relative offset was estimated between these two vertical datum realizations at the North and South Islands. For this purpose, the geometric geoid/quasigeoid heights (obtained from GPS and newly adjusted levelling data) were compared to the regional gravimetric geoid/quasigeoid models. Moreover, the oceanographic and geodetic models of mean dynamic topography (MDT) were used to assess the relative offset between these two vertical datum realizations through the analysis of regional spatial variations of mean sea level (MSL).

Four different regional gravimetric geoid/quasigeoid solutions were used to analyse the systematic trend in the newly adjusted levelling networks, namely the KTH geoid model and the NZGeoid-2009, BEM and OTG12 quasigeoid models. The KTH geoid model was compiled by Abdalla and Tenzer (2011) based on applying the method developed at the Royal Institute of Technology (KTH) in Stockholm. The theoretical and numerical aspects of the KTH method can be found in Sjöberg (1984, 1991, 2003a, 2003b, 2003c, 2003d, 2004). The BEM quasigeoid model of New Zealand was compiled by Tenzer et al. (2012a) by applying the boundary element method (BEM). This method was developed byČunderlík et al. (2008) and Čunderlík and Mikula (2009). The OTG12 gravimetric quasigeoid model was determined by Abdalla and Tenzer (2012) based on applying the discretized integral-equation approach. We also used the official quasigeoid model of New Zealand—NZGeoid2009, which was prepared by Claessens, et al. (2011).

6.1　Geodetic vertical datum in New Zealand— historical background

The geodetic vertical reference system at the North and South Islands of New Zealand was realized by 12 major LVDs based on precise levelling from 11 different tide gauges. The LVD Dunedin-Bluff 1960 was defined by fixing the heights of two levelling benchmarks from the LVDs Dunedin 1958 and BLUFF 1955 instead of using the tide gauge as the origin. Moreover, additional LVDs were established for surveying purposes throughout the country based on precise levelling from tide gauges or connecting to existing levelling networks. The LVDs were defined in the system of (approximate) normal-orthometric heights. The cumulative normal-orthometric correction to levelled height differences was defined based on the GRS67 normal gravity field parameters and computed approximately using a truncated form of the GRS67 normal-orthometric correction formula (Rapp, 1961). Since LVDs were referenced to the local MSL determined based on the analysis of tide-gauge records, large discrepancies exist between individual LVDs. These discrepancies in New Zealand reach up to a few decimetres at the inter-connecting levelling benchmarks of which heights were defined in two LVDs (Amos and Featherstone, 2009).

The unification of LVDs can be done either by a joint adjustment of local levelling networks or by a determination of the gravimetric geoid/quasigeoid model and a subsequent combination of gravimetric solution with GPS-levelling data. Amos and Featherstone (2009) argued that the quality of levelling data is too low to be used for a joint readjustment of the levelling networks at the North and South Islands of New Zealand due to several reasons, among others the realization of levelling networks over a period of several decades, vertical tectonic motions, and loss of large number of levelling benchmarks. Therefore, they developed and applied the iterative gravimetric approach to unify the LVDs in New Zealand. This method facilitated an iterative determination of the regional gravimetric quasigeoid model and its comparison with the geometric quasigeoid model determined using GPS-levelling data on each LVD. The results of this method were provided in terms of the average LVD offsets relative to the regional quasigeoid model. Amos and Featherstone (2009) applied this method to estimate the LVD offsets relative to the first official national quasigeoid model—New Zealand Quasigeoid 2005 (NZGeoid2005), see also Amos (2007). The LVD offsets obtained from comparing the gravimetric and geometric quasigeoid models were used to apply additional reductions to gravity observations within each LVD. Amos and Featherstone (2009) showed that the solution converges after only three iterations. According to their results, the estimated offsets of 13 LVDs at the North, South, and Stewart Islands range between 24 and 58 cm with the estimated standard deviation of about 8 cm. The New Zealand Quasigeoid 2009 (NZGeoid2009) is the currently adopted official national quasigeoid model for New Zealand (Claessens et al., 2009, 2011). NZGeoid2009 is provided to users on a 1×1 arc-min geographical grid over New Zealand and its continental shelf. NZGeoid2009 was determined again by applying the iterative gravimetric approach. The main difference between compiling NZGeoid2005 and NZGeoid2009 was due to using different GGMs for modelling the reference gravity field, NZGeoid2005 was computed using EGM96 (Lemoine et al., 1998) and EGM2008 (Pavlis et al., 2008) was used for NZGeoid2009.

6.2   Unification of vertical datum

Since not only the levelling data, but essentially also the gravity data in New Zealand are sparse and very irregularly distributed, it was advisable to use all available geodetic information to unify the LVDs. For this reason we jointly (re)adjusted the levelling networks at the North and South Islands of New Zealand and analysed their relations with respect to the regional gravimetric geoid/quasigeoid models and WHS (Tenzer et al. , 2011b; Abdalla and Tenzer, 2012). This procedure was realized in several processing steps comprising the estimation of the LVD offsets, the height conversion, the joint adjustment of levelling networks and finding their relation with respects to WHS. The methodology and results are summarized below.

6. 2. 1　LVD offsets

The cumulative normal to normal-orthometric height correction was first applied to convert the original normal-orthometric heights to the normal heights. A more detailed discussion of this step is postponed until Section 6. 2. 3. A similar method was used by Filmer, et al. (2010) for the conversion of the normal-orthometric to normal heights in the Australian Height Datum. They used EGM2008 to reconstruct the observed gravity data at the levelling benchmarks of the Australian National Levelling Network. Equivalently, the orthometric heights should be converted to the normal heights before applying the geopotential-value approach to estimate the LVD offsets. Burša et al. (1999), for instance, applied the geoid-to-quasigeoid correction to estimate the average offset of the North American Vertical Datum 1988 (NAVD88) which was realized in the system of Helmert's orthometric heights.

The geopotential-value approach was further applied to determine the offsets of major LVDs at the North and South Islands. The principle of this method is based on estimating the

45

geopotential differences at GPS-levelling points of a particular LVD relative to the adopted geoidal geopotential value W_0. This approach was developed by Burke et al. (1996) and later applied by Burša et al. (1999, 2001) to estimate the average offsets of major LVDs in Europe, North America, and Australia. A similar method was used by Grafarend and Ardalan (1997) and Ardalan and Grafarend (1999) to calculate the LVD offsets in Baltic countries. Burša et al. (1997, 2007) estimated the geoidal geopotential value of $W_0 = 62,636,856 \pm 0.5$ m^{2}s^{-2}. This value was later adopted by the International Astronomical Union (IAU). It is worth mentioning that different values of W_0 were reported by Sanchez (2007) and Dayoub et al. (2012). Sanchez (2007) determined the value of W_0 using different MSL models and different GGMs showing that the choice of MSL and GGM is unimportant for estimating W_0, while the latitude domain of the altimetry-derived MSL models plays a major role. The value of W_0, estimated by Sanchez (2007), differs by 2.5 m^{2}s^{-2} from the value adopted by IAU. In a more recent study, Dayoub et al. (2012) reviewed previous studies using various methods and datasets. They confirmed the conclusions of Sanchez (2007), but reported and recommended a different value of $W_0 = 62,636,854.2 \pm 0.5$ m^{2}s^{-2} and established that the dependency of W_0 on the latitude domain is merely due to the MDT.

The geopotential-value approach utilises Molodensky's concept of the normal heights according to which the normal gravity potential U evaluated at the telluroid equals the actual gravity potential W at the topographic surface (cf. Molodensky et al., 1960)

$$U(H^N) = W(h) . \tag{6.1}$$

In practice, however, the condition in Eq. (6.1) does not

hold due to the fact that the geopotential value $W_{0, LVD}$ at the tide gauge, used as the reference of LVD, generally differs from the geoidal geopotential value W_0. The geopotential difference $\delta W_{0, LVD}$ between the values of W_0 and $W_{0, LVD}$ is then computed as (Burša et al., 1999)

$$\delta W_{0, LVD} = W_0 - W_{0, LVD} = U(H^N) - W(h) . \qquad (6.2)$$

From Eq. (6.2), the LVD offset evaluated at the GPS-levelling point is defined as (ibid.)

$$\delta H_{0, LVD} = \frac{\delta W_{0, LVD}}{\bar{\gamma}} = \frac{U(H^N) - W(h)}{\bar{\gamma}}. \qquad (6.3)$$

The gravity potential W in Eq. (6.3) is computed at the topographic surface point using the GGM coefficients, and the normal gravity potential U is computed at the telluroid using, for instance, Somigliana's formula (Somigliana, 1929; see also Heiskanen and Moritz, 1967, Eqs. 2-62).

Tenzer et al. (2011a, 2011b) applied this method to estimate the LVDs offsets in New Zealand relative to WHS. They used the EGM2008 coefficients complete to degree 2160 of spherical harmonics (in the tide-free system) to compute the potential values at the topographic surface. The geopotential differences were computed at the GPS-levelling points and then averaged for each LVD. The configuration of the GPS-levelling testing network in New Zealand (consisting of 2, 127 points) attributed to 14 LVDs is shown in Fig. 6. 1. The geodetic heights were defined in the New Zealand Geodetic Datum 2000 (NZGD2000; GRS80 reference ellipsoid). The NZGD2000 is aligned to the International Terrestrial Reference Frame 1996 (ITRF1996) at the reference epoch of January 1st 2000 (Blick et al., 2005). The estimated average offsets of 14 LVDs, summarised in Table 6. 1, vary from 1 cm (Wellington 1953 LVD) to 37 cm (One Tree Point 1964 LVD).

Table 6. 1　**The estimated average offsets and their uncertainties of 14 LVDs in New Zealand (Tenzer et al. , 2011b). The LVD offsets are taken relative to WHS (defined based on $W_0 =$ 62, 636, 856 $m^2 s^{-2}$).**

	LVD	LVD offset[cm]
North Island	Auckland 1946	12 ±4
	Gisborne 1926	10 ±4
	Moturiki 1953	19 ±9
	Napier 1962	24 ±6
	One Tree Point 1964	37 ±5
	Taranaki 1970	12 ±6
	Wellington 1953	1 ±2
South Island	Bluff 1955	17 ±4
	Deep Cove 1960	30 ±5
	Dunedin-Bluff 1960	23 ±7
	Dunedin 1958	7 ±18
	Lyttelton 1937	13 ±11
	Nelson 1955	20 ±9
	Tarakohe 1982	23 ±3

The LVDs Wellington 1953 and Dunedin 1958 have the smallest average offsets relative to WHS. The estimated average offsets of these two LVDs are 1±2 cm (for the LVD Wellington 1953) and 7±18 cm (for the LVD Dunedin 1958). Therefore, the tide gauges in Wellington and Dunedin were chosen as the origins for a definition of the new normal-orthometric heights at the North and South Islands.

6. 2. 2　Levelling network adjustment

The national precise levelling networks at the North and

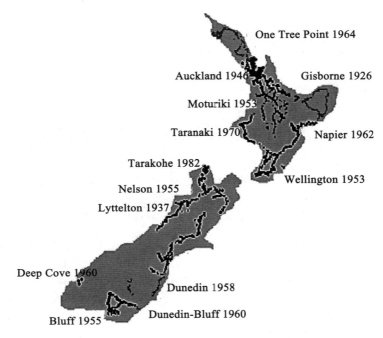

One Tree Point 1964

Auckland 1946

Gisborne 1926

Moturiki 1953

Taranaki 1970

Napier 1962

Tarakohe 1982

Nelson 1955

Wellington 1953

Lyttelton 1937

Deep Cove 1960

Dunedin 1958

Bluff 1955

Dunedin-Bluff 1960

Fig. 6. 1   The GPS-levelling testing network in New Zealand
attributed to 14 LVDs

South Islands of New Zealand(Gilliland, 1987) comprised in total
10, 150 benchmarks (5, 967 levelling benchmarks at the North
Island and 4, 183 levelling benchmarks at the South Island). The
configuration of the levelling networks is shown in Fig. 6. 2. The
whole network consists of 14 LVDs. As seen in Fig. 6. 2, large
parts of the South Island are not sufficiently covered by levelling
profiles along the mountainous regions of the Southern Alps. Over
most of the North Island, the coverage of levelling networks is
much better, except for some irregularities in the mountainous
regions of the central and lower North Island.

Since the normal gravity values at the surface points along
levelling lines were calculated using the parameters of the GRS67
reference ellipsoid, we firstly recomputed the cumulative normal-

49

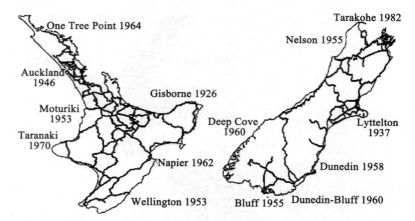

Fig. 6. 2   The levelling networks at the North and South Islands of New Zealand attributed to 14 LVDs (Auckland 1946, Gisborne 1926, Moturiki 1953, Napier 1962, One Tree Point 1964, Taranaki 1970, and Wellington 1953 at the North Island; Bluff 1955, Deep Cove 1960, Dunedin-Bluff 1960, Dunedin 1958, Lyttelton 1937, Nelson 1955, and Tarakohe 1982 at the South Island).

orthometric correction using the normal gravity field parameters of the GRS80 reference ellipsoid (Moritz, 2000). The test results confirmed the finding of Filmer et al. (2010) that the differences between the values of the cumulative normal-orthometric correction computed for the GRS67 and GRS80 normal gravity field parameters are completely negligible.

Despite the normal-orthometric-corrected loop closures are not independent of the levelling route taken (Featherstone and Kuhn, 2006), the accurate computation of the normal correction to levelled height differences was restricted (in the absence of observed gravity data along levelling lines) by the cumulative effect of the EGM2008 commission and omission errors, especially in the mountainous regions with large spatial gravity and elevation gradients. The test results indicated that the errors

Table 6. 2   **Statistics of the least-squares residuals between the measured and adjusted normal-orthometric-corrected height differences at the levelling benchmarks at the North and South Islands of New Zealand.**

LVD	Min[cm]	Max[cm]	STD[cm]
Auckland 1946	−0. 8	0. 8	0. 1
Gisborne 1926	−1. 0	0. 9	0. 2
Moturiki 1953	−0. 8	0. 01	0. 2
Napier 1962	−0. 7	0. 6	0. 1
One Tree Point 1964	−2. 5	2. 6	0. 3
Taranaki 1970	−0. 6	0. 7	0. 1
Wellington 1953	−1. 9	1. 9	0. 2
North Island	**−2. 5**	**2. 6**	**0. 2**

LVD	Min[cm]	Max[cm]	STD[cm]
Bluff 1955	−0. 5	0. 5	0. 1
Deep Cove 1960	−0. 2	1. 4	0. 1
Dunedin-Bluff 1960	−0. 2	0. 03	0. 1
Dunedin 1958	−1. 0	1. 0	0. 2
Lyttelton 1937	−1. 0	1. 1	0. 2
Nelson 1955	−1. 3	1. 3	0. 2
Tarakohe 1982	−0. 3	0. 3	0. 1
South Island	**−1. 3**	**1. 4**	**0. 2**

in computed values of the cumulative normal to normal-orthometric height correction reach up to a few centimetres. Therefore, the observation equations in the joint adjustment of levelling networks were formed for the normal-orthometric-corrected loop closures, while disregarding the holonomity

51

property (meaning, among other things, that the normal or orthometric corrected loop closures are equal zero independently on the levelling route; cf. Sansò and Vaníček, 2006).

The least-squares residuals between the measured and adjusted normal-orthometric-corrected height differences at the levelling benchmarks are shown in Fig. 6. 3 and the histograms of residuals are plotted in Fig. 6. 4. The statistics of the residuals at the levelling benchmarks of individual LVDs at the North and South Islands are summarised in Table 6. 2.

The results of the joint levelling adjustment revealed a good quality of levelling data by means of the residuals between the measured and adjusted normal-orthometric-corrected height differences at the levelling benchmarks. The standard deviation (STD) of the least-squares residuals is 2 mm for the whole country. At the South Island the residuals range within ±1. 3 cm, and they are between −2. 5 and 2. 6 cm at the North Island. The most precise levelling networks were attributed to the LVDs Tarakohe 1982 (where the residuals are within ± 0. 3 cm) and Dunedin-Bluff 1960 (where the residuals are between − 0. 2 and 0. 03 cm) at the South Island. The largest residuals are found at the levelling networks of the LVDs One Tree Point 1964 (between −2. 5 and 2. 6 cm) and Wellington 1953 (within ±1. 9 cm) at the North Island. As seen in Table 6. 2, the levelling networks of the LVDs Moturiki 1953 (at the North Island) and Deep Cove 1960 (at the South Island) have a systematic trend (mostly either positive or negative) in the residuals. A possible reason is due to the location of these LVDs in the mountainous regions with large horizontal elevation gradients.

The new normal-orthometric heights of the levelling networks at the South and North Islands were computed from the heights of the fixed tide-gauge reference benchmarks and the adjusted

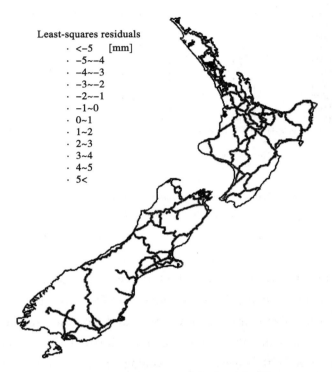

Least-squares residuals
· <-5   [mm]
· -5~-4
· -4~-3
· -3~-2
· -2~-1
· -1~0
· 0~1
· 1~2
· 2~3
· 3~4
· 4~5
· 5<

Fig. 6. 3   Results of the joint adjustment of the local levelling networks at the
North and South Islands: The least-squares residuals between the
measured and adjusted normal-orthometric-corrected height differences
at the levelling benchmarks.

normal-orthometric-corrected height differences. The differences between the newly determined and original normal-orthometric heights of the levelling benchmarks in New Zealand are shown in Fig. 6. 5 and statistics aresummarized in Table 6. 3. These differences are between – 26. 5 and 23. 4 cm at the levelling benchmarks of the North Island and between –21. 6 and 6. 5 cm at the South Island. The individual comparison of the differences between the newly determined and original normal-orthometric

53

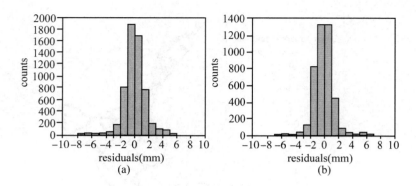

Fig. 6. 4   *The histograms of the least-squares residuals between the measured and adjusted normal-orthometric-corrected height differences at the levelling benchmarks of*: (*a*) *the North Island and* (*b*) *the South Island.*

heights at 14 LVDs revealed that the levelling networks of the LVDs Auckland 1946, Gisborne 1926, and One Tree Point 1964 at the North Island are mainly attributed to the spatial MSL variations around the coast of the North Island while the contribution of the systematic levelling errors is much smaller. The averaged values of the relative offsets between these LVDs taken with respect to the LVD Wellington 1953 are: 2. 3 cm for the LVD Auckland 1946, 0. 1 cm for the LVD Gisborne 1926, and −3. 0 cm for the LVD One Tree Point 1964. Similarly, the levelling networks of the LVDs Bluff 1955, Deep Cove 1960, Dunedin-Bluff 1960, Dunedin 1958, and Tarakohe 1982 at the South Island can be attributed to the spatial MSL variations around the coast of the South Island. The averaged values of the relative offsets between these LVDs taken with respect to the LVD Dunedin 1958 are: 6. 3 cm for the LVDs Bluff 1955 and Deep Cove 1960, 6. 0 cm for the LVD Dunedin-Bluff 1960, and − 11. 6 cm for the LVD Tarakohe 1982. These relative LVD offsets only partially agree with the

Table 6. 3 **Statistics of differences between the original and newly determined normal-orthometric heights of levelling benchmarks at the North and South Islands of New Zealand.**

LVD	Min[cm]	Max[cm]	Mean[cm]	RMS[cm]
Auckland 1946	0. 5	0. 6	0. 5	0. 1
Gisborne 1926	−2. 2	−1. 4	−1. 7	0. 2
Moturiki 1953	1. 5	23. 4	7. 6	2. 5
Napier 1962	1. 7	12. 4	15. 0	1. 0
One Tree Point 1964	−4. 8	−4. 7	−4. 8	0. 1
Taranaki 1970	−26. 5	−0. 1	−12. 9	6. 1
Wellington 1953	−5. 9	1. 0	−1. 8	1. 7
North Island	**−26. 5**	**23. 4**	**3. 2**	**6. 9**

LVD	Min[cm]	Max[cm]	Mean[cm]	RMS[cm]
Bluff 1955	6. 4	6. 5	6. 4	0. 1
Deep Cove 1960	6. 4	6. 5	6. 4	0. 1
Dunedin-Bluff 1960	6. 0	6. 1	6. 1	0. 1
Dunedin 1958	−0. 4	0. 8	0. 1	0. 3
Lyttelton 1937	−10. 7	−5. 8	−7. 4	0. 6
Nelson 1955	−21. 6	−13. 3	−18. 0	1. 7
Tarakohe 1982	−11. 5	−11. 4	−11. 5	0. 1
South Island	**−21. 6**	**6. 5**	**−6. 5**	**7. 4**

global principal pattern of the increasing MDT due to a typical south-north horizontal temperature gradient. The reasons are more likely due to local spatial MDT irregularities attributed to the coastal configuration, the geometry of the ocean bottom relief, and the seawater circulation around the coast of New Zealand which is dominated by the East Auckland Current, East Cape Current, Westland Current and D'Urville Current. The individual comparison of the differences between the original and newly

55

determined normal-orthometric heights at 14 LVDs in Table 6. 3 also indicates the presence of large discrepancies possibly caused by the systematic levelling errors at the North Island of the LVDs Moturiki 1953, Napier 1962, Taranaki 1970, and Wellington 1953 and the South Island of the LVDs Lyttelton 1937 and Nelson 1955.

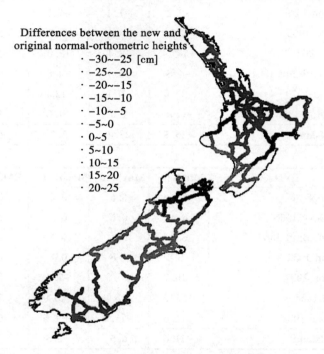

Differences between the new and
original normal-orthometric heights
· −30~−25 [cm]
· −25~−20
· −20~−15
· −15~−10
· −10~−5
· −5~0
· 0~5
· 5~10
· 10~15
· 15~20
· 20~25

Fig. 6. 5   The differences between the original and newly determined normal-
orthometric heights of the levelling benchmarks at the North and South
Islands of New Zealand.

The GPS-levelling data and the EGM2008 coefficients were finally used to estimate the average offsets of the jointly adjusted levelling networks at the North and South Islands with respect to WHS, which was defined by the adopted geoidal geopotential value of $W_0 = 62, 636, 856 \pm 0.5 \ m^2 s^{-2}$ (Burša et al. , 2007). For this purpose, we converted the newly determined normal-orthometric heights into the normal heights (according to the procedure described in Section 6.2.3). The estimated offsets for the jointly adjusted levelling networks at the North and South Islands were found to be 10.6 and 27.5 cm respectively. The resulting normal-orthometric heights were obtained from the adjusted normal-orthometric heights by applying these estimated offsets.

6.2.3 Height conversion

The cumulative normal to normal-orthometric height correction was computed using the levelling and gravity anomaly data according to the following expression (Tenzer et al., 2011b)

$$\delta H^{N,N\text{-}O} = H^N - H^{N\text{-}O} = \frac{1}{\gamma} \sum_i g_i \Delta H_i - \frac{1}{\gamma} \sum_i \left[\gamma_{0,i} + \frac{\partial \gamma}{\partial h} H_i^{N\text{-}O} \right] \Delta H_i$$

$$= \frac{1}{\gamma} \sum_i \Delta g_i \Delta H_i , \qquad\qquad (6.4)$$

where g_i are the observed gravity values along levelling lines, $H_i^{N\text{-}O}$ are the normal-orthometric heights, $\gamma_{0,i}$ are the normal gravity values computed at the reference ellipsoid, and $\partial \gamma / \partial h$ is the normal linear gravity gradient. As seen from Eq. (6.4), the normal to normal-orthometric height correction along levelling lines was calculated cumulatively by summing up the product of the levelled height differences ΔH_i and the values of the gravity anomaly Δg_i.

The gravity anomalies at the surface points along levelling lines were calculated using the EGM2008 coefficients complete to degree 2,160 of spherical harmonics in the tide-free system. For a definition of tidal systems we refer readers, for instance, to Vatrt (1999). The practical aspects of geodetic data conversion in New Zealand between different permanent tidal systems were discussed in detail by Tenzet et al. (2011a). The computed values of the

EGM2008 gravity anomalies at the levelling benchmarks vary from −157. 1 to 115. 4 mGal (the gravity anomalies vary from −84. 2 to 115. 4 mGal at the levelling benchmarks of the South Island, and between −157. 1 and 102. 5 mGal at the levelling benchmarks of the North Island).

The values of the normal to normal-orthometric height correction computed at the levelling benchmarks in New Zealand are shown in Fig. 6. 6, and their statistics are given in Table 6. 4. At the North Island this correction varies from −4. 9 to 10. 7 cm, while it varies between −2. 6 and 5. 7 cm at the South Island. The mostly positive values of this correction have their maxima at the levelling lines crossing the mountainous regions of the central

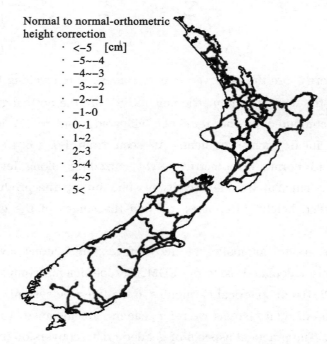

Fig. 6. 6   The normal to normal-orthometric height correction computed at the levelling benchmarks in New Zealand

Table 6. 4　**Statistics of the normal to normal-orthometric height correction computed at the levelling benchmarks at the North and South Islands of New Zealand.**

LVD	Min[cm]	Max[cm]	Mean[cm]	RMS[cm]
Auckland 1946	0. 0	1. 4	0. 3	0. 3
Gisborne 1926	−3. 3	2. 6	−1. 6	1. 4
Moturiki 1953	−0. 5	10. 7	2. 8	3. 2
Napier 1962	−4. 9	2. 9	−0. 9	1. 6
One Tree Point 1964	−0. 2	2. 3	0. 3	0. 4
Taranaki 1970	−0. 1	1. 5	0. 6	0. 4
Wellington 1953	−0. 5	2. 2	0. 7	0. 6
North Island	**−4. 9**	**10. 7**	**0. 3**	**0. 7**

LVD	Min[cm]	Max[cm]	Mean[cm]	RMS[cm]
Bluff 1955	−0. 1	0. 2	0. 0	0. 1
Deep Cove 1960	−0. 1	0. 1	0. 0	0. 01
Dunedin-Bluff 1960	−0. 1	1. 6	0. 6	0. 3
Dunedin 1958	−0. 1	2. 2	0. 0	0. 3
Lyttelton 1937	−2. 6	5. 7	−0. 5	1. 2
Nelson 1955	−0. 1	5. 6	1. 4	0. 8
Tarakohe 1982	0. 0	5. 1	1. 0	1. 5
South Island	**−2. 6**	**5. 7**	**0. 4**	**0. 4**

North Island (LVD Moturiki 1953) and the upper South Island (LVD Lyttelton 1937) . The largest negative values of this correction are detected along the central levelling segment of the

LVD Napier 1962. Large negative values of this correction were also found at levelling lines along the west coastline of the South Island (LVD Lyttelton 1937) and the east coast of the North Island (LVD Gisborne 1926). Both, the correction maxima and minima are situated at the levelling segments with the largest horizontal gravity and terrain elevation gradients.

The computation of the approximate geoid-to-quasigeoid correction was done according to Eq. (4. 14) using the simple planar Bouguer gravity anomalies and topographic heights of levelling points. The computed values of the simple planar Bouguer gravity anomalies at the levelling benchmarks are between −166. 0 and 91. 4 mGal (from −111. 5 to 91. 4 mGal at the levelling benchmarks of the South Island, and from − 166. 0 to 73. 1 mGal at the North Island). It is worth mentioning here that a smoothing effect of the simple Bouguer planar reduction on gravity data is not significant due to a low spatial resolution of the EGM2008 gravity anomalies.

The values of the geoid-to-quasigeoid correction computed at the levelling benchmarks in New Zealand are shown in Fig. 6. 7, and their statistics aresummarized in Table 6. 5. This correction is mostly positive. At the North Island it varies from − 1. 5 to 9. 0 cm, while it varies between −2. 5 and 6. 5 cm at the South Island. The maxima of this correction are at the levelling benchmarks located at high elevations in the mountainous regions of the Southern Alps (at the South Island) and the central North Island. The presence of the largest (positive) values of the geoid-to-quasigeoid correction was explained by prevailing negative values of the Bouguer gravity anomalies in the mountainous regions. Over the flat regions with low elevations, this correction is mostly negative due to mainly positive values of the Bouguer gravity anomalies in these areas.

Table 6. 5   **Statistics of the geoid-to-quasigeoid correction computed at the levelling benchmarks at the North and South Islands of New Zealand.**

LVD	Min[cm]	Max[cm]	Mean[cm]	RMS[cm]
Auckland 1946	−0. 8	−0. 0	−0. 1	−0. 1
Gisborne 1926	−0. 6	3. 7	0. 0	1. 5
Moturiki 1953	−1. 5	6. 2	0. 6	1. 5
Napier 1962	−1. 1	9. 0	1. 0	1. 6
One Tree Point 1964	−1. 2	−0. 0	−0. 1	−0. 2
Taranaki 1970	−0. 4	1. 7	0. 3	0. 3
Wellington 1953	−1. 0	1. 7	0. 2	0. 3
North Island	**−1. 5**	**9. 0**	**0. 2**	**1. 1**

LVD	Min[cm]	Max[cm]	Mean[cm]	RMS[cm]
Bluff 1955	−0. 2	0. 0	−0. 1	0. 0
Deep Cove 1960	0. 4	0. 8	0. 6	0. 1
Dunedin-Bluff 1960	−0. 6	1. 1	0. 1	0. 4
Dunedin 1958	−1. 1	4. 1	−0. 1	0. 1
Lyttelton 1937	−2. 5	6. 5	0. 2	0. 5
Nelson 1955	−1. 4	2. 2	0. 1	0. 3
Tarakohe 1982	−0. 9	2. 4	−0. 0	0. 1
South Island	**−2. 5**	**6. 5**	**0. 2**	**0. 6**

The differences between the Helmert orthometric heights and the normal-orthometric heights at the levelling benchmarks in New Zealand are shown in Fig. 6. 8, and the statistics are given in Table 6. 6. At the North Island, these differences vary from −3. 2 to 13. 0 cm, and between −2. 9 and 7. 9 cm at the South Island.

Table 6. 6 **Statistics of the differences between the Helmert orthometric heights and the normal-orthometric heights at the levelling benchmarks at the North and South Islands of New Zealand.**

LVD	Min[cm]	Max[cm]	Mean[cm]	RMS[cm]
Auckland 1946	−0. 8	0. 8	0. 0	0. 2
Gisborne 1926	−3. 0	3. 7	−0. 1	1. 3
Moturiki 1953	−1. 5	13. 0	1. 5	3. 1
Napier 1962	−3. 2	9. 0	0. 5	1. 3
One Tree Point 1964	−1. 2	1. 3	0. 0	−0. 2
Taranaki 1970	−0. 4	2. 2	0. 4	0. 6
Wellington 1953	−0. 9	3. 4	0. 4	0. 6
North Island	**−3. 2**	**13. 0**	**1. 2**	**2. 7**

LVD	Min[cm]	Max[cm]	Mean[cm]	RMS[cm]
Bluff 1955	−0. 2	0. 0	0. 0	0. 05
Deep Cove 1960	0. 4	0. 8	0. 5	0. 3
Dunedin-Bluff 1960	−0. 6	2. 3	0. 3	0. 5
Dunedin 1958	−1. 1	4. 3	0. 3	0. 7
Lyttelton 1937	−2. 9	7. 9	0. 1	1. 0
Nelson 1955	−1. 4	5. 9	0. 8	1. 0
Tarakohe 1982	−0. 9	7. 6	0. 5	1. 4
South Island	**−2. 9**	**7. 9**	**0. 4**	**0. 6**

The accuracy of computed corrections depends mainly on the errors of input gravity data, whereas the errors due to the inaccuracy of the levelled heights are by comparison much smaller. Since the precise levelling in New Zealand was realised without observing the gravity and the GNS Science gravity database comprised only about 40,000 gravity measurements in New Zealand with an irregular spatial distribution, the gravity

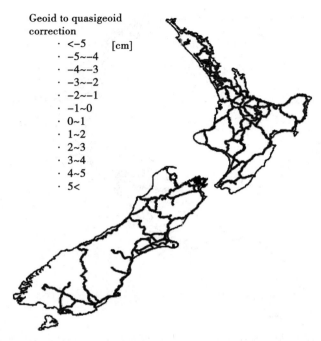

Geoid to quasigeoid
correction

· <−5   [cm]
· −5~−4
· −4~−3
· −3~−2
· −2~−1
· −1~0
· 0~1
· 1~2
· 2~3
· 3~4
· 4~5
· 5<

*Fig. 6.7   The geoid-to-quasigeoid correction computed at the levelling benchmarks
in New Zealand.*

anomalies at the levelling benchmarks were calculated using
EGM2008. The comparison of the EGM2008-derived and observed
(from the GNS Science Database) gravity anomalies at gravity
points revealed large discrepancies up to several dozens of
miligals. A simple error analysis showed that, for example, the
error of 10 mGal in the values of the gravity anomaly causes the
error of the computed geoid-to-quasigeoid correction of 1 cm at
the elevation of 1,000 m, while only 1 mm error at the elevation
of 100 m. On the other hand, large errors were expected in the
computed values of the normal to normal-orthometric height
correction due to the cumulative contribution of the EGM2008
commission and omission errors. The numerical analysis revealed

63

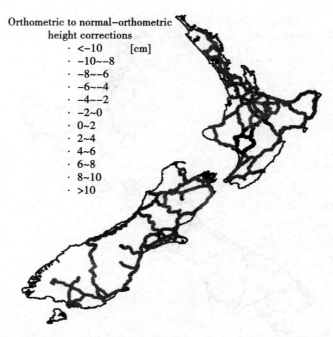

Orthometric to normal–orthometric
height corrections
- · <–10   [cm]
- · –10~–8
- · –8~–6
- · –6~–4
- · –4~–2
- · –2~0
- · 0~2
- · 2~4
- · 4~6
- · 6~8
- · 8~10
- · >10

Fig. 6. 8   The differences between the Helmert orthometric heights and the normal-orthometric heights at the levelling benchmarks in New Zealand.

large systematic differences up to a few centimetres in the values of the normal to normal-orthometric correction computed along different levelling routes, particularly along the levelling segments in mountainous regions.

6. 2. 4   Analysis of systematic trends

The newly-adjusted normal-orthometric heights (corrected for the average offsets relative to W_0) and converted into the systems of (Molodensky's) normal and (Helmert's) orthometric heights (in Section 6. 2. 2) were used together with the geodetic heights to calculate the geometric geoid/quasigeoid heights at the 1,452 GPS-levelling benchmarks (772 at the North Island, and 680 at the South Island) . These geometric geoid/quasigeoid heights were compared with the respective gravimetric geoid/ quasigeoid heights of the NZGeoid2009, KTH, BEM and OTG12 regional models. The values of the geoid/quasigeoid heights at the locations of GPS-levelling testing points were determined from grid values by applying a linear interpolation; NZGeoid2009 is provided on a 1×1 arc-min grid, and the KTH, BEM and OTG12 models were compiled on a 2×2 arc-min grid.

Five different MDT solutions offshore New Zealand were also used to investigate the MSL differences between the tide gauges in Dunedin and Wellington. These MDT solutions included the oceanographic models CARS2009 (Ridgway et al., 2002) and ECCO2 (Menemenlis et al., 2008) as well as the geodetic MDT model DTU10 (Andersen, 2010). In addition, we used two geodetic MDT solutions which were derived from the datasets of sea surface topography DOT. DNSC08 (Tapley et al., 2003) and CLS11 (Scharroo, 2011; Schaeffer et al., 2011) after subtracting the EGM2008 marine geoid heights. The MDT solutions CARS2009, ECCO2, DTU10, DOT. DNSC08-EGM2008 and CLS11-EGM2008 at the study area are shown in Fig. 6.9, the corresponding statistics are summarized in Table 6.7.

The oceanographic MDT models CARS2009 and ECCO2 have a low resolution and spatial coverage especially in coastal seas (Fig. 6.9). CARS2009 and ECCO2 have also a significantly smaller range of values within the study area than the geodetic MDT models; the range of CARS2009 is 81 cm while ECCO2 has only 51 cm. The MDT range of DTU10 and CLS11-EGM2008 is 100 cm. The DOT. DNSC08-EGM2008 has the largest MDT variations within the study area at a range of 110 cm. All investigated MDT solutions showed a similar pattern with a prevailing zonal trend of increasing MDT towards tropical seas due to latitudinal thermal gradient. Regional anomalous features associated with the configuration of ocean currents (dominated by the influence of

65

Tasman and Sub-Antarctic Fronts) were also recognized; for more details see Tenzer et al. (2012a). The most significant regional feature taken into consideration in our analysis was a slightly higher MSL in Wellington compared to Dunedin's coastal sea.

The newly-determined orthometric/normal heights at the GPS-levelling testing network were compared with four regional geoid/quasigeoid models. The geometric geoid heights were calculated from the NZGD2000 geodetic heights by subtracting the orthometric heights. The geometric quasigeoid heights (i. e., height anomalies) were obtained from the NZGD2000 geodetic heights by subtracting the normal heights. The differences between the geometric and gravimetric geoid heights were computed for the KTH geoid model. The corresponding differences between the geometric and gravimetric quasigeoid heights were computed for the NZGeoid2009, BEM and OTG12 quasigeoid models. The same analysis was realized using the original levelling data defined in 14 LVDs. The differences between the geometric and gravimetric geoid/quasigeoid heights for the newly adjusted and original levelling data are plotted in Figs. 6.10 and 6.11 respectively.

Table 6.7 **Statistics of the MDT models: CARS2009, ECCO2, DTU10, DOT. DNSC08-EGM2008 and CLS11-EGM2008 within the study area shown in Fig. 6.9.**

MDT	Min[cm]	Max[cm]	Mean[cm]	STD[cm]
CARS2009	164	245	216	17
ECCO2	−24	47	11	20
DTU10	−7	93	58	19
DOT. DNSC08-EGM08	−49	61	27	19
CLS11-EGM08	−31	69	29	18

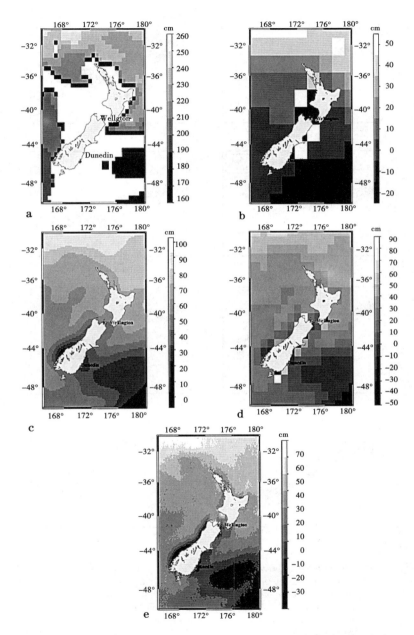

Fig. 6.9 The oceanographic and geodetic MDT solutions: (a) CARS2009, (b) ECCO2,
(c) DTU10, (d) DOT. DNSC08-EGM2008, and (e) CLS11-EGM2008

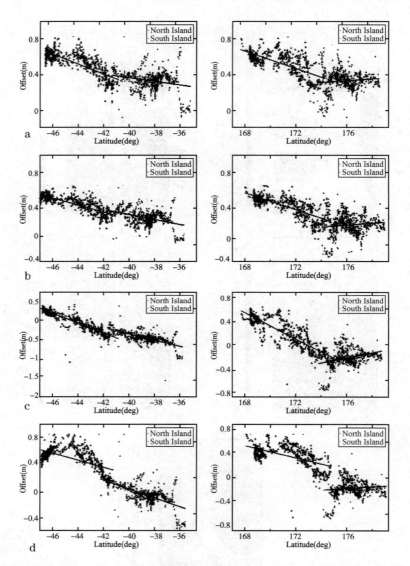

Fig. 6. 10   The differences between the geometric and gravimetric geoid/quasigeoid heights along meridional (left panels) and parallel (right panels) profiles computed using: (a) KTH, (b) NZGeoid2009, (c) BEM, and (d) OTG12. The geometric geoid/quasigeoid heights were computed based on the jointly adjusted levelling data at the South and North Islands (and corrected for the average offsets relative to W_0). The linear regression analysis was applied to fit the differences by a linear trend function for each island.

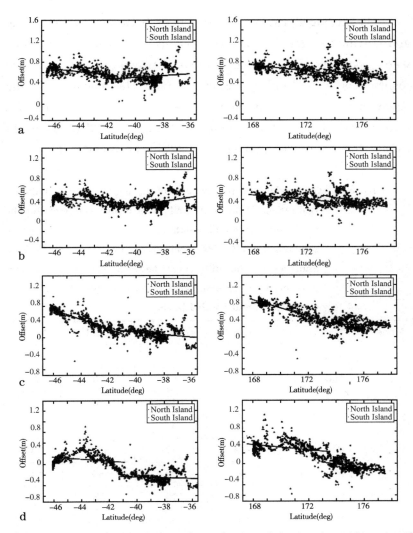

Fig. 6.11   *The differences between the geometric and gravimetric geoid/quasigeoid heights along meridional (left panels) and parallel (right panels) profiles computed using: (a) KTH, (b) NZGeoid2009, (c) BEM, and (d) OTG12. The geometric geoid/quasigeoid heights were computed using the original levelling data attributed to 14 LVDs. The linear regression analysis was applied to fit the differences by a linear trend function for each island.*

The averaged values of differences between the geometric and gravimetric geoid/quasigeoid heights at GPS-levelling points were used to estimate a relative offset between the vertical datum realizations at the North and South Islands. This was done for the newly adjusted levelling data. Moreover, these computations were repeated using the original levelling data attributed to 14 LVDs. The results are summarized in Tables 6. 8 and 6. 9.

Table 6. 8   **The values of the relative offset between vertical datum realizations at the North and South Islands computed for the newly adjusted levelling data using four regional gravimetric solutions (KTH, NZGeoid2009, BEM, and OTG12). Statistics of the differences between the geometric and gravimetric geoid/quasigeoid heights computed individually at the North and South Islands.**

Model		KTH	NZGeoid09	BEM	OTG12
North Island	Min[cm]	−70	0	−75	−46
	Max[cm]	118	91	70	85
	STD[cm]	13	12	16	14
	Mean[cm]	40	42	−17	5
South Island	Min[cm]	−49	17	−84	−44
	Max[cm]	100	99	86	121
	STD[cm]	14	13	23	18
	Mean[cm]	65	63	23	61
Relative Offset[cm]		−25	−21	−40	−56

The comparison of the regional gravimetric solutions with the newly adjusted levelling data (corrected for the average offsets relative to W_0) revealed large discrepancies (see Fig. 6. 10). For KTH and NZGeoid2009, the differences between the geometric and gravimetric geoid/quasigeoid heights are mainly positive with the largest values at the lower South Island and the corresponding smallest differences at the upper North Island. The BEM gravimetric solution is not significantly biased from levelling data. The largest absolute differences (exceeding ~ 50 cm) are seen at the lower South Island and at the upper North Island. The differences between GPS-levelling data and OTG12 reach maxima (~1 m) in the central South Island, while the largest negative differences (~ – 40 cm) are found at the upper North Island. Elsewhere in the North Island these differences are typically within ±20 cm.

Table 6. 9   **The values of the relative offset between vertical datum realizations at the North and South Islands computed for the original levelling data (defined in 14 LVDs) using four regional gravimetric solutions (KTH, NZGeoid2009, BEM, and OTG12). Statistics of the differences between the geometric and gravimetric geoid/quasigeoid heights computed individually at the North and South Islands.**

Model		KTH	NZGeoid09	BEM	OTG12
North Island	Min [cm]	−45	−36	−109	−80
	Max [cm]	118	140	131	85
	STD [cm]	15	15	15	14
	Mean[cm]	46	48	−11	11
South Island	Min [cm]	−56	−55	−90	−51
	Max [cm]	93	94	81	126
	STD [cm]	12	11	22	16
	Mean[cm]	57	55	23	53
Relative Offset[cm]		−11	−7	−34	−42

71

As seen from these results, the KTH and NZGeoid2009 gravimetric solutions are biased with respect to the GPS-levelling results. In addition, the presence of a large systematic trend across New Zealand is seen in all four gravimetric solutions. The BEM and OTG12 gravimetric solutions have the largest systematic discrepancies (reaching up to ~1 m). The misfits of the KTH and NZGeoid2009 gravimetric solutions with respect to GPS-levelling data are more similar; the range of geoid/quasigeoid heights differences is ~40 cm (for KTH) and ~50 cm (for NZGeoid2009). These large discrepancies can be explained by the systematic errors within either gravimetric solutions or levelling data (or both). On the other hand, the presence of relative offset between the vertical datum realizations at two islands is less obvious. The character of the geoid/quasigeoid heights differences at the GPS-levelling points at the upper South Island and the lower North Island (plotted in Fig. 6.10) is relatively smooth without any significant (inter-island) discontinuity. Similarly, the linear regression fit of the differences between the geometric and gravimetric geoid/quasigeoid heights computed separately for each island does not exhibit any significant misfit. This is evident especially for the KTH and NZGeoid2009 solutions. The misfit of the linear regression trends between both islands is less than 10 cm while the corresponding misfit for BEM and OTG12 is ~20 cm.

In order to better understand these large discrepancies which were found among the regional gravimetric solutions and GPS-levelling data, the analysis was done for the corresponding differences computed using the original levelling data. As seen in Fig. 6.11, the systematic trend and bias in the computed differences between the GPS-levelling and gravimetric results is now much less pronounced. The regional gravimetric solutions thus better agree with the original levelling data defined in 14 LVDs. The KTH and NZGeoid2009 models are systematically biased (~50 cm) from the GPS-levelling results. The BEM quasigeoid model is more consistent with the GPS-levelling results. The systematic discrepancies are, in this case, seen particularly at the South Island, where the range of differences is ~40 cm. The OTG12 quasigeoid model has, on the other hand, significantly different fit with the GPS-levelling data at the South and North Islands.

The relative offsets computed when taking into consideration only the GPS-levelling points close to tide gauges are similar to that found when using all GPS-levelling points. The estimated relative offsets are 17 cm (for KTH) and 26 cm (for NZGeoid2009). The relative offsets for BEM and OTG12 are about two times larger, namely 52 cm (for BEM) and 44 cm (for OTG12).

The estimation of a relative offset between vertical datum realizations at the North and South Islands could be biased by the presence of systematic trend which is seen in plotted differences between the geometric and gravimetric geoid/quasigeoid heights. Therefore, we estimated the relative offset between the newly established vertical datum realizations at the North and South Islands from these differences but taken only at GPS-levelling points in the vicinity of tide gauges in Wellington and Dunedin. The results are summarized in Table 6. 10.

The character of systematic distortions within the levelling networks and regional gravimetric solutions was analysed based on their comparison with the EGM2008 quasigeoid model (computed using the spherical harmonic coefficients complete to degree/order of 2160). The differences between the GPS-levelling and EGM2008 quasigeoid heights for the original and newly adjusted levelling data were plotted in Fig. 6. 12; statistics of these differences are given in Table 6. 11. The differences between the regional gravimetric solutions and EGM2008 are

73

plotted in Fig. 6. 13; statistics of these differences are given in Table 6. 12. The geoid-to-quasigeoid correction was applied to the KTH geoid heights for the comparison with EGM2008.

Table 6. 10 **The values of the relative offset between vertical datum realizations at the North and South Islands computed for the newly adjusted levelling data using four regional gravimetric solutions (KTH, NZGeoid2009, BEM, and OTG12). The differences between the geometric and gravimetric geoid/ quasigeoid heights were averaged from the values taken at GPS-levelling points in the vicinity of tide gauges in Wellington and Dunedin.**

Gravimetric model		KTH	NZGeoid09	BEM	OTG12
North Island	Mean[cm]	56	54	10	31
	STD [cm]	2	3	2	3
South Island	Mean[cm]	73	80	62	75
	STD [cm]	6	5	6	4
Relative Offset [cm]		−17	−26	−52	−44

The comparison of GPS-levelling data with EGM2008 (Fig. 6. 12) revealed a slightly better agreement of the newly adjusted levelling data in terms of existing systematic bias. The mean of differences is 1 and 3 cm for the North and South Islands respectively. The corresponding mean values of differences obtained when using the original levelling data are −5 cm (for the North Island) and 5 cm (for the South Island). The STD of differences computed using the newly adjusted levelling data is 11 cm for both islands. A better STD of differences of 8 cm was found for the original levelling data at the South Island while the STD of differences at the North Island is 14 cm.

74

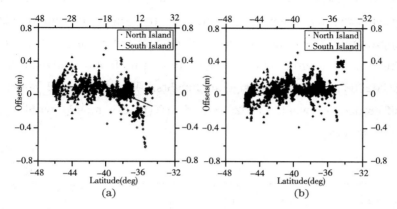

Fig. 6. 12   *The differences between the geometric and EGM2008 (gravimetric) quasigeoid heights computed using: (a) the original levelling data attributed to 14 LVDs , and (b) the newly adjusted levelling data (and corrected for the average offsets relative to W_0).*

Table 6. 11   **Statistics of the differences between the GPS-levelling and EGM2008 quasigeoid heights computed using the original and newly adjusted levelling data shown in Fig. 6. 12.**

	North Island		South Island	
	Original levelling data	Newly adjusted levelling data	Original levelling data	Newly adjusted levelling data
Min[cm]	−68	−46	−33	−40
Max[cm]	51	44	40	32
Mean[cm]	−5	1	5	−3
STD[cm]	14	11	8	11

As seen in Fig. 6. 13, all four regional gravimetric solutions are systematically displaced from EGM2008. The discrepancies

75

between EGM2008 and the KTH and NZGeoid2009 regional gravimetric solutions are similar to the range of differences approximately within - 30 to - 80 cm. A much larger range of differences (within 40 to - 80 cm) was found between the EGM2008 and BEM quasigeoid heights. The quasigeoid heights differences between EGM2008 and OTG12 are mostly within 20 and -80 cm.

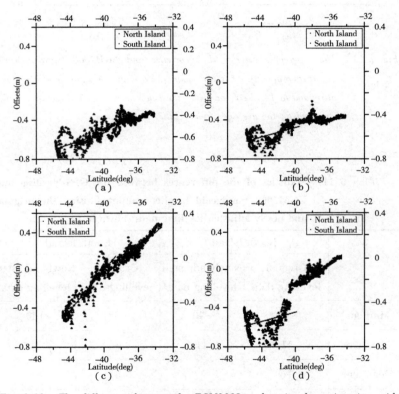

Fig. 6. 13   The differences between the EGM2008 and regional gravimetric geoid/ quasigeoid models: (a) KTH, (b) NZGeoid2009, (c) BEM, and (d) OTG12.

Table 6. 12 **Statistics of the differences between the EGM2008 quasigeoid model and the regional gravimetric solutions (KTH, NZGeoid2009, BEM, and OTG12) shown in Fig. 6. 13.**

	KTH		NZGeoid09		BEM		OTG12	
	North Island	South Island	North Island	South Island	North Island	South Island	North Island	South Island
Min [cm]	−62	−114	−51	−108	−12	−94	−26	−141
Max [cm]	−27	−36	−25	−43	41	21	10	−17
Mean [cm]	−42	−65	−44	−59	16	−28	−6	−58
STD [cm]	6	11	3	9	10	20	7	17

Finally, the MSL offsetwas estimated between tide gauges in Wellington and Dunedin using the MDT solutions CARS2009, ECCO2, DTU10, DOT. DNSC08-EGM2008 and CLS11-EGM2008. The MSL values were calculated by extrapolating the MDT grid values in the vicinity of these two tide gauges. The results are summarized in Table 6. 13.

The analysis of MSL in the vicinity of tide gauges in Wellington and Dunedin based on five MDT models showed that the MSL offsets between these two tide gauges are between 18 and 25 cm when taking into consideration only the results of the geodetic models DTU10, DOT. DNSC08-EGM2008, and CLS11-EGM2008 (Table 6. 13). The MSL offset of 29 cm was found in the oceanographic model CARS2009 while only 1 cm for ECCO2. The representative MSL offset obtained by averaging these results is ~ 19 cm. When disregarding ECCO2 model (which is more likely unrealistically small), the average MSL offset increases to ~ 24 cm.

Table 6. 13　**The values of the MSL offset between the tide gauges（TG）in Wellington and Dunedin computed using the MDT solutions: CARS2009, ECCO2, DTU10, DOT. DNSC08-EGM2008, and CLS11-EGM2008.**

Model		CARS 2009	ECCO2	DTU10	DOT. DNSC08 -EGM08	CLS11 -EGM2008
TG Wellington (North Island)	Min [cm]	201	−3	50	19	12
	Max [cm]	216	−3	63	31	37
	STD [cm]	1	0	3	3	6
	Mean[cm]	206	−2	55	23	20
TG Dunedin (South Island)	Min [cm]	176	−8	29	−5	−5
	Max [cm]	177	−2	42	6	13
	STD [cm]	1	1	2	2	3
	Mean[cm]	177	−3	31	−2	2
Relative Offset[cm]		29	1	24	25	18

6. 3　Discussion

The levelling and normal gravity data were used to readjust the local levelling networks at the North and South Islands of New Zealand while fixing the heights of the tide gauges in Dunedin and Wellington. the geopotential-value approach was then applied to estimate the average offsets of the jointly-adjusted levelling networks relative to WHS using the GPS-levelling data and EGM2008. Since the geopotential-value approach is based on Molodensky's theory, the newly adjusted normal-orthometric heights were for this purpose converted to the normal heights based on applying the cumulative normal to normal-orthometric height correction. In the absence of observed gravity data along levelling lines, the gravity anomaly values were generated from EGM2008.

The results of the joint levelling adjustment revealed that the STD of least-squares residuals of the normal-orthometric-corrected height differences is 2 mm in New Zealand. The residuals between the levelling benchmarks are within ±1.3 cm at the South Island and between −2.5 and 2.6 cm at the North Island.

The comparison of the newly determined normal-orthometric heights (defined with respect to the tide gauges in Dunedin and Wellington) with the original ones (defined individually in 14 LVDs) confirmed the presence of large offsets between individual LVDsand systematic levelling errors.

The results of the height conversion revealed that the differences between the normal and normal-orthometric heights vary from −4.9 to 10.7 cm. The absolute maxima of this correction are along the levelling lines crossing mountainous regions with the largest horizontal gravity and terrain elevation gradients. The differences between the Helmert orthometric heights and the normal heights vary between −2.5 and 9.0 cm. The maxima of this correction are at the highest elevations of levelling benchmarks inthe mountainous regions with large negative values of the simple Bouguer gravity anomaly.

The average offsets of GPS-levelling points at the North and South Islands were estimated to be 10.6 and 27.5 cm respectively. The uncertainty in these estimated values was mainly attributed to the EGM2008 commission and omission errors, the existing systematic distortions of the levelling networks, the errors in GPS solutions and geocentric reference frame realisation, the effects of sea level variations, tectonic and other vertical movements, the definition of MSL based on short

79

term tide-gauge records, and the levelling network realisation over different time epochs.

Two, principally different, methods were applied to investigate a relative offset between the vertical datum realizations at the North and South Islands. First, the GPS-levelling data were compared with the regional gravimetric solutions. The regional gravimetric solutions were also compared with EGM2008. Abdalla and Tenzer (2012) conducted a comparative analysis of the recent satellite-only GRACE and GOCE models with EGM2008 (using spherical harmonics up to degree 250). They have shown show that EGM2008 has the best regional agreement with GPS-levelling data in terms of STD of residuals. This might be explained by the fact that the GPS-levelling data were used directly for the compilation of EGM2008. Various oceanographic and geodetic MDT solutions were further used to estimate the MSL offset between the tide gauges in Wellington and Dunedin.

The analysis of MDT models (CARS2009, DTU10, DOT. DNSC08-EGM2008, and CLS11-EGM2008 while disregarding ECCO2) revealed that the average MSL offset between tide gauges in Wellington and Dunedin is ~24 cm. This value is ~7 cm larger than the estimated relative offset (of 16.9 cm) between the jointly adjusted levelling networks at the North and South Islands obtained from their comparison with WHS (cf. Tenzer et al., 2011b). The estimated value of MSL offset from the analysis of MDT models is, however, to a large extent affected by a low accuracy of altimetry data in coastal seas as well as uncertainties of the regional marine geoid model. A better agreement of this MSL offset was found with relative offsets obtained from comparing the KTH and NZGeoid2009 gravimetric solutions with GPS-levelling data. The values of relative offset between the vertical datum realizations and regional gravimetric solutions are

80

25 cm (for KTH) and 21 cm (for NZGeiod2009) . The corresponding relative offsets of 17 cm (for KTH) and 26 cm (for NZGeiod2009) were obtained when averaging the geoid/ quasigeoid heights differences only at GPS-levelling points in the vicinity of tide gauges (in order to avoid a possible influence of systematic errors within levelling networks and/or regional gravimetric solutions). All these estimates of relative offsets are about two times smaller than the corresponding values obtained from the comparison of the BEM and OTG12 regional gravimetric solutions with GPS-levelling data. These findings indicate that the estimates for BEM and OTG12 are more likely unrealistically large. These two gravimetric solutions have also the largest misfits with all solutions used for the analysis in this study.

The results of numerical analysis revealed that the relative offset between the vertical datum realizations at the North and South Islands was to a large extent eliminated by correcting the heights of levelling benchmarks for the offset with respect to WHS (after applying the geopotential-value approach) . This was evident from plotted values of the differences between the GPS-levelling and regional gravimetric solutions in Fig. 6. 13. These differences did not exhibit any significant discontinuity between both islands. This was also confirmed by comparing the newly adjusted levelling data with EGM2008. From these results we can conclude that, despite the existence of large systematic errors within levelling networks and regional gravimetric solutions, the new vertical datum realizations at the North and South Islands of New Zealand are relatively closely interconnected. The expected uncertainty in the relative inter-island offset of ~ 10 cm is at the level of errors of the used GPS and levelling data, the regional gravimetric geoid/quasigeoid solutions and EGM2008.

7. Topographic Effect on
Gravity and Geoid

The average density of $2,670$ kg m^{-3} is typically attributed to the upper continental crust in geological and gravity surveys, geophysical explorations, gravimetric geoid modelling, compilation of regional gravity maps, and other applications. Although this density value is widely used, its origin remains partially obscure. Woollard (1966) suggested that this density was used for the first time by Hayford and Bowie (1912). In reviewing several studies seeking a representative mean density from various rock type formations, Hinze (2003) argued that this value was used earlier by Hayford (1909) for the gravity reduction. Hayford (1909) referred to Harkness (1891) who averaged five published values of surface rock density. The value of Harkness (1891) of $2,670$ kg m^{-3} was confirmed later, for instance, by Gibb (1968) who estimated the mean density of the surface rocks in a significant portion of the Canadian Precambrian shield from over $2,000$ individual measurements. Woollard (1962) examined more than $1,000$ rock samples and estimated that the mean basement (crystalline) rock density is about $2,740$ kg m^{-3}. Subrahmanyam and Verma (1981) determined that crystalline rocks in low-grade metamorphic terranes in India have the mean density of $2,750$ kg m^{-3}, while $2,850$ kg m^{-3} in high-grade metamorphic terranes.

It is often assumed that when disregarding water bodies, the variation of actual topographic density is mostly within ± 300 kg m^{-3} around the average value of $2,670$ kg m^{-3} (e. g. , Martinec, 1998). Therefore, the effect of anomalous topographic density should amount to about 10% of the total topographic effect (cf. Huang et al. , 2001). However, larger topographic mass density variations (20% ~ 30%) are encountered in some other parts of the world (e. g. , Tziavos and Featherstone, 2000). The mass density lateral variations are documented to generate centimetres to decimetres changes on the geoid (Vaníček et al. , 1995; Tenzer and Vaníček, 2003; Hwang and Hsiao 2003; Allister and Featherstone, 2001). Based on the results conducted in a high-elevation and rugged part of the Canadian Rocky Mountains, Tenzer, et al. (2005) demonstrated that this effect ranges from -7 to 2 cm, while the total effect of topography (including lateral density variation) on the geoid-to-quasigeoid correction varies between -86.5 and 0.1 cm. They also showed that the contribution of the density heterogeneities below the geoid surface was within a range of -8 to 44 cm.

These studies indicate that the assumptionof a constant topographic density can introduce large inaccuracy in the orthometric heights, the geoid surface and consequently also in estimated values of the geoid-to-quasigeoid correction. To investigate the effect of the lateral topographic density, the rock density model of New Zealand was compiled using the digital geological data and rock density samples. The rock density model was then used to estimate the effect of variable (lateral) topographic density on the gravity and geoid. It is worth mentioning that until now a little is known about the stratigraphic (vertical) density distribution due to the lack of borehole density samples, except for some limited information on the density increase due to compaction within sedimentary basins.

83

7. 1 Geological setup of New Zealand

The geological composition of New Zealand's land surface is characterized mainly by Cenozoic, particularly Quaternary, sedimentary and pyroclastic volcanic deposits. Many of these rocks were deposited beneath the sea adjacent to the present or past plate boundaries and later uplifted and juxtaposed by tectonic movement. The hard greywacke sandstone and mudstone of Mesozoic age form large areas of the South Island and the Southern Alps east of the Alpine Fault. Greywacke basement also forms the axial ranges of the southern and eastern North Island and eastern Northland. In central Otago, schist predominates at the surface and has originated from metamorphism of the Mesozoic greywacke sedimentary rock. A great proportion of the southern half of the North Island is formed of soft Neogene rocks, particularly sandstone and mudstone. Limestone is widespread throughout the North and South Islands and is generally thin, although thicker formations are also widely presented. Large volcanic areas occur in the central and northern North Island. Intrusive igneous rocks dominated by granite, diorite, granodiorite and tonalite, but including ultramafic rocks mostly occur in Nelson, Westland, Fiordland and Stewart Island. For a more detailed description of the geological composition of New Zealand with the detailed literature review of relevant geological studies we refer readers to Tenzer eft al. (2011c).

7. 2 Rock density model of New Zealand

The digital geological maps QMAP, the rock density measurements from the national rock catalogue PETLAB, and supplementary geological sources were used to compile the digital rock density model of New Zealand (Tenzer et al. , 2011c).

The QMAP (Quarter-million MAP) database produced by GNS Science provides national geological map coverage at 1 : 250,000 in printed and digital form using ESRI's ArcGIS Geographic Information System (GIS) software. The database was derived from numerous sources such as older published and unpublished geological maps, mining company reports, petroleum exploration reports, university theses, unpublished research reports, and data collected from new field work. The QMAP database contains thematic layers with rich attributes that describe various features of a geological map. The most relevant for this study were the geological unit polygons that define the extent of mapping units. The units mapped are generally the shallowest rock unit more than 5 to 10 metres thick. A map of main rock types in New Zealand was generated from the digital QMAP database. The QMAP database identifies 123 main rock types, not including areas of water (lakes) and ice (glaciers and snowfields). The generalised geological map of New Zealand consisting of 18 broad categories with place-names is shown in Fig. 7. 1.

The PETLAB is the rock catalogue and geo-analytical database of New Zealand (Mortimer, 2005). It is operated by GNS Science in collaboration with the geology departments of New Zealand's universities. The database contains locations, descriptions and analyses of rock and mineral samples collected throughout onshore and offshore New Zealand and Antarctica. Information was sourced from journal articles, theses, and open file reports. PETLAB contains 157,363 sample records from which 40,588 have analytical data. Wet density measurements compiled from many sources cover 89 rock types collected at

85

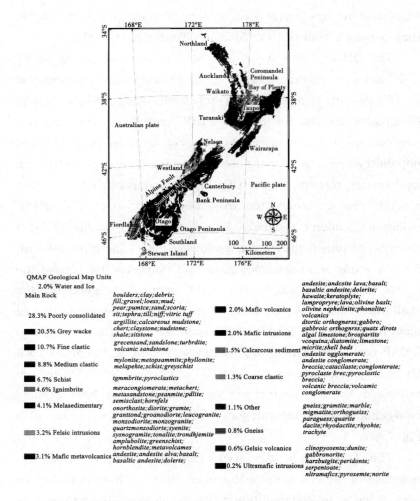

Fig. 7. 1   The geological map of broad groups of main rock types in New Zealand generated from the digital QMAP geological map database (Tenzer et al. , 2011c).

9,256 locations in New Zealand. The location map of rock density samples from the PETLAB rock catalogue is shown in Fig. 7. 2.

The compilation of the rock density model from the vector GIS map of the main rock types was realized in three processing

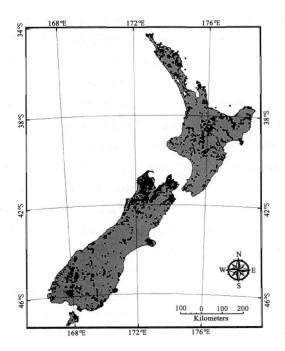

Fig. 7. 2   The location map of PETLAB rock density samples in New Zealand.

steps. First, the densities were assigned to the main rock types of the QMAP database. Since the main rock type applies to one or more geological mapping units, the assigned density was assumed to represent the geological mapping units also. This results in a vector GIS map of main rock type densities. The rock density model was then obtained from the vector map after applying the data discretisation and aggregation procedures.

The primary source of information used to allocate the representative densities for 123 main rock types of the QMAP database was the PETLAB rock catalogue. The PETLAB rock catalogue provides 8,933 rock density measurements for the 56 main rock types in the QMAP database. For *in situ* near subsurface values, the wet density measurement is the most

87

appropriate. The wet density values for different rock types throughout New Zealand were extracted from the PETLAB rock catalogue and tabulated. The average rock densities and complementary statistics are summarized in Table 7.1. The representative value of density for each main rock type was computed by averaging over all PETLAB samples of the same rock type collected throughout New Zealand. Variations in density samples taken for the same rock type often exceed $1,000$ kg m^{-3} and can reach $1,630$ kg m^{-3} (volcanic breccias). This indicates the practical restrictions in allocating rock density values objectively. The densities of the remaining 67 main rock types, for which the information from the PETLAB rock catalogue was not found, were allocated according to available sources from the literature or by assuming similarity or synonymity with other rock types. The list of rock density values and the references to relevant sources are given in Table 7.2. A density of 920 kg m^{-3} and $1,000$ kg m^{-3} was attributed to ice and water respectively. The mean density of the main rock types varies between 900kg m^{-3} (loess) and $3,300$kg m^{-3} (dunite).

The vector map of the main rock type densities was discretized on a 5×5 arc-sec equal angular grid of geographical coordinates. The 5×5 arc-sec grid of the main rock type densities was then aggregated into the 1×1 arc-min spatial resolution digital density model using a mean operator.

The rock density model at 1×1 arc-min spatial resolution for New Zealand is shown in Fig. 7.3. The surface rock densities vary between 900 and $3,300$ kg m^{-3}. The average rock density (without lakes and glaciers) was found from a spatial averaging of the 5×5 arc-sec grid data. It indicates an average surface rock density of about $2,440$ kg m^{-3}, with the standard deviation of 280 kg m^{-3}. When accounting for glaciers and lakes (roughly 2% of the total area of New Zealand), the average surface density decreases to about $2,415$ kg m^{-3}.

Table 7. 1 **Statistics of 56 rock type densities from the PETLAB rock catalogue.**

Rock Type	Mean [kg/m^3]	STD [kg/m^3]	Max [kg/m^3]	Min [kg/m^3]	Number of Samples	Area [%]
AMPHIBOLITE	2,892	115	3,100	2,630	62	0.3
ANDESITE	2,565	170	2,990	1,560	418	0.8
ARGILLITE	2,691	133	3,160	2,000	251	0.3
BASALT	2,768	162	3,060	1,780	340	1.7
BRECCIA	2,291	295	3,000	1,540	118	0.3
CHERT	2,564	162	2,740	2,240	11	< 0.05
CLAY	2,067	171	2,450	1,920	9	0.1
CLAYSTONE	2,067	235	2,420	1,520	15	0.1
COAL	1,712	461	2,100	1,130	5	< 0.05
CONGLOMERATE	2,570	159	3,000	2,110	118	0.9
DACITE	2,402	175	2,700	1,940	79	0.1
DIATOMITE	1,528	141	1,720	1,390	4	< 0.05
DIORITE	2,797	119	3,160	2,430	186	1.4
DOLERITE	2,749	146	3,040	2,360	68	< 0.05
GABBRO	2,884	147	3,340	2,260	236	0.4
GNEISS	2,812	179	3,150	1,830	149	0.5
GRANITE	2,640	77	2,940	2,330	288	2.3
GRANODIORITE	2,681	70	2,940	2,530	53	0.6
GRANULITE	2,765	332	3,000	2,530	2	< 0.05
GRAVEL	2,309	266	2,580	1,870	9	21.4
GREENSAND	2,365	168	2,520	2,210	4	< 0.05

Continued

Rock Type	Mean [kg/m^3]	STD [kg/m^3]	Max [kg/m^3]	Min [kg/m^3]	Number of Samples	Area [%]
GREENSCHIST	2,923	133	3,130	2,620	15	0.2
GREYWACKE	2,639	100	2,940	2,090	469	20.5
HORNFELS	2,800	143	3,080	2,590	22	< 0.05
IGNIMBRITE	2,125	247	2,620	1,240	916	4.6
LAMPROPHYRE	2,910	177	3,390	2,650	15	< 0.05
LAVA	2,680	93	2,820	2,540	6	< 0.05
LIGNITE	1,390	204	1,670	1,210	6	< 0.05
LIMESTONE	2,484	211	3,010	1,890	156	< 0.05
MARBLE	2,716	117	3,140	2,510	22	< 0.05
METAVOLCANIC	2,955	124	3,150	2,760	11	< 0.05
MUDSTONE	2,204	301	2,870	1,320	734	8.1
MYLONITE	2,757	91	2,930	2,620	11	0.1
PERIDOTITE	3,093	225	3,340	2,340	118	< 0.05
PHONOLITE	2,536	70	2,630	2,470	5	< 0.05
PUMICE	1,719	310	2,230	1,150	29	0.1
PYROCLASTIC	1,986	512	2,350	1,130	5	< 0.05
PYROXENITE	3,122	218	3,330	2,240	25	< 0.05
QUARTZITE	2,612	73	2,780	2,490	29	0.1
RHYOLITE	2,207	225	2,740	1,360	704	0.5
SAND	2,048	351	3,220	1,690	27	3.2
SANDSTONE	2,463	266	3,000	1,510	968	8.2
SCHIST	2,732	115	3,100	2,120	419	5.8
SERPENTINITE	2,634	219	3,270	2,240	87	0.1
SHALE	2,335	190	2,730	1,610	193	< 0.05

Continued

Rock Type	Mean [kg/m^3]	STD [kg/m^3]	Max [kg/m^3]	Min [kg/m^3]	Number of Samples	Area [%]
SILT	1,979	140	2,160	1,720	14	0.2
SILTSTONE	2,347	283	2,880	1,360	471	2.0
SINTER	2,510	14	2,520	2,500	2	< 0.05
SPILITE	2,863	121	3,090	2,500	82	0.1
SYENITE	2,719	145	2,860	2,440	15	< 0.05
TEPHRA	1,637	98	1,750	1,580	3	< 0.05
TRACHYTE	2,591	182	2,950	2,170	31	< 0.05
TUFF	2,113	289	2,940	1,410	723	0.2
ULTRAMAFIC	3,288	517	4,130	2,750	5	< 0.05
VOLCANICS	2,362	471	3,090	1,480	110	< 0.05
VOLCANIC BRECCIA	2,195	358	2,950	1,320	60	< 0.05

Table 7.2 **The densities of 67 main rock types allocated according to supplementary sources.**

Rock Type	Area [%]	Density [kg/m^3]	Source and Comment
LOESS	0.2	900	Johnson and Lorenz (2000)
PEAT	0.7	1,040	Schon (1,996)
PYROCLASTIC BREC	< 0.05	1,600	Hall et al. (1999)
MUD	1.3	1,910	Clark (1966)
VITRIC TUFF	< 0.05	2,113	The adopted density the same as for Tuff
LAPILLI TUFF	< 0.05	2,113	The adopted density the same as for Tuff

Continued

Rock Type	Area [%]	Density [kg/m³]	Source and Comment
CALCAREOUS MUDSTONE	< 0.05	2,200	The adopted density the same as for Mudstone
CATACLASITE	< 0.05	2,291	The adopted density the same as for Breccia
LEUCOGRANITE	< 0.05	2,291	Annen and Scaillet (2006)
TILL	0.1	2,310	Balco and Stone (2003)
RHYODACITE	< 0.05	2,350	Hildreth et al. (2004)
HORNBLENDITE	< 0.05	2,370	Clark (1966)
BOULDERS	0.1	2,400	Nott (2003)
DEBRIS	0.9	2,400	The adopted density the same as for Boulders
FILL	< 0.05	2,400	The adopted density the same as for Boulders
TURBIDITE	0.5	2,410	Average density of Sandstone and Siltstone
VOLCANIC SANDSTONE	0.2	2,460	The adopted density the same as for Sandstone
HAWAIITE	0.3	2,470	Carmichael (1982)
ALGAL LIMESTONE	< 0.05	2,480	The adopted density the same as for Limestone
COQUINA	< 0.05	2,480	The adopted density the same as for Limestone
MICRITE	< 0.05	2,480	The adopted density the same as for Limestone
SHELL BEDS	< 0.05	2,482	The adopted density the same as for Limestone (2484)
KERATOPHYRE	< 0.05	2,500	Morrow and Lockner (2001)

Continued

Rock Type	Area [%]	Density [kg/m³]	Source and Comment
PORPHYRY	< 0.05	2,550	Andrew (1995)
SCORIA	< 0.05	2,550	Tamari et al. (2005)
METACHERT	< 0.05	2,560	The adopted density the same as for Chert
PELITE	< 0.05	2,560	Pettijohn (1975)
ANDESITE AGGL./CONGL.	< 0.05	2,570	The same as Andesite and Conglomerate
METACONGLOMERATE	< 0.05	2,570	The adopted density the same as for Conglomerate
VOLCANIC CONGL.	< 0.05	2,570	The adopted density the same as for Conglomerate
MONZODIORITE	0.1	2,580	Llambias et al. (1977)
SYENOGRANITE	< 0.05	2,610	Gaal et al. (1981)
ANDESITE LAVA	< 0.05	2,630	Hildreth et al. (2004)
ORTHOGNEISS	0.1	2,630	Giacomini et al. (2009)
PSAMMITE	0.2	2,639	The adopted density the same as for Greywacke
TRONDHJEMITE	< 0.05	2,640	Carmichael (1982)
MELANGE	0.5	2,660	Kimura et al. (2001)
BASALTIC ANDESITE	< 0.05	2,670	Averagedensity of Basalt and Andesite
BROKEN FORMATION	0.2	2,670	No information found
CALC-SILICATE	< 0.05	2,670	No information found
CLINOPYROXENITE	< 0.05	2,670	No information found
SEMISCHIST	0.8	2,686	Average density of Schist and Greywacke

93

Continued

Rock Type	Area [%]	Density [kg/m³]	Source and Comment
BIOSPARITE	< 0.05	2,690	Allaby (1999)
MONZOGRANITE	< 0.05	2,690	Oliveira et al. (2008)
DIORITIC ORTHOGNEISS	0.1	2,700	Pechinig et al. (2005)
GABBROIC ORTHOGNEISS	< 0.05	2,700	Pechinig et al. (2005)
GRANITOID	< 0.05	2,700	Rao et al. (2008)
METASANDSTONE	2.0	2,700	Carmichael (1982)
OLIVINE BASALT	< 0.05	2,700	Arrnient et al. (1991)
PARAGNEISS	0.2	2,700	Samalikova (1983)
TRAVERTINE	< 0.05	2,710	Russell and Pellant (1981)
GREYSCHIST	0.9	2,730	The adopted density the same as for Schist
PHYLLONITE	< 0.05	2,740	Wibberley and Mc Caig (2000)
METAPELITE	1.1	2,750	Dyda (1994)
QUARTZ MONZODIORITE	< 0.05	2,770	Clark (1966)
METAPSAMMITE	< 0.05	2,800	Clark (1966)
TONALITE	0.2	2,800	Nettleton et al. (1969)
QUARTZ DIORITE	0.1	2,810	Clark (1966)
ANORTHOSITE	< 0.05	2,810	Clark (1966)
MIGMATITE	< 0.05	2,812	The adopted density the same as for Gneiss
GABBRONORITE	< 0.05	2,970	Christensen and Mooney (1995)
LIMBURGITE	< 0.05	2,970	Vankova and Kropacek (1974)
NORITE	< 0.05	2,980	Clark (1966)
OLIVINE NEPHELINITE	< 0.05	3,150	Martinkova et al. (2000)

Continued

Rock Type	Area [%]	Density [kg/m³]	Source and Comment
HARZBURGITE	0. 1	3,200	Arafin et al. (2008)
DUNITE	< 0. 05	3,300	Bullen (1966)

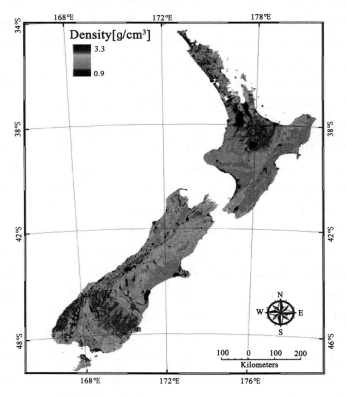

Fig. 7. 3   The rock density model of New Zealand.

The geographical configuration of surface rock densities (Fig. 7. 3) resembles the major geological composition of New Zealand (Tenzer et al. , 2011c). The locations of rock units with

95

lowest densities are correlated with volcanic areas of the central North Island and with large areas of volcanic deposits in the North Island. The northern, southern and eastern parts of the North Island consist of heavier sedimentary rock formations. In the South Island, large areas of the central Otago, Fiordland, and along the Alpine Fault have higher rock densities due to the dominance of schist, greywacke and intrusive rocks. The locations of lower rock densities in the South Island correspond to the locations of sedimentary rocks of sandstone and mudstone and unconsolidated gravel, sand and mud.

The analysis of the 9,256 wet density samples in the PETLAB database revealed that the average density of these samples is 2,450 kg m^{-3}, with the standard deviation of 360 kg m^{-3}. This value incorporates some sample bias in the PETLAB dataset, for example, unconsolidated sediments such as gravel, sand, clay and silt that form 28% of the land area are under-represented with only 0.6% of the measured density data. The rock densities vary from 1,130 to 5,480 kg m^{-3}, with 90% of the values ranging between 1,780 and 2,930 kg m^{-3}(see the histogram in Fig. 7.4).

The average density 2,450 kg m^{-3}estimated from the PETLAB rock density samples very closely agrees with the value of 2,440 kg m^{-3} obtained from the rock density model. These values of the average rock density in New Zealand are smaller than the average density of 2,670 kg m^{-3} defined based on the average value of crystalline and granitic rock formations (Hinze, 2003). We explained this by the fact that large areas of New Zealand are capped by Cenozoic, particularly Quaternary, sedimentary and pyroclastic volcanic deposits. The average rock density of 2,336 kg m^{-3} in the North Island reflects the predominance of un-metamorphosed sedimentary rock, tephra and ignimbrite. The average rock density of 2,514 kg m^{-3} in the South Island, on the

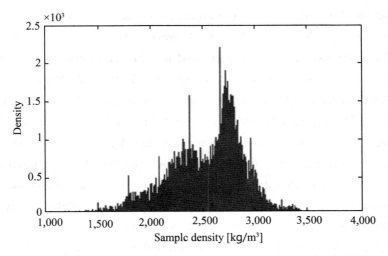

Fig. 7. 4   The histogram of 9, 256 wet density samples from the PETLAB rock
catalogue.

other hand, reflects the influence of more common igneous intrusions and metamorphosed sedimentary rock, including schist and gneiss.

7. 3   Gravitational effect of topographic density variations

The method of Hayford and Bowie (1912) and Hammer (1939) was applied to calculate the gravitational effect of topography based on subdividing the surface integration domain into zones and compartments. The gravitational contributions of individual compartments were calculated according to Mikuška et al. (2006) and using the height and density for each compartment interpolated from digital terrain and density models. The surface integration domain was subdivided into the near and

97

distant zones. The near zone was taken up to 166. 7 km of the spherical distance from the computation point (i. e. outer limit of the Hayford Zone O). The 1×1 arc-sec DTM of New Zealand (Columbus et al. , 2011) was used for the integration within the near zone, and the 30×30 arc-sec SRTM30PLUS _ V5. 0 global elevation data (Becker et al. , 2009) were used for the distant zone. The topographic gravity corrections for the average density (of $2,670$ kg m^{-3}) and the anomalous density were computed at the 1×1 arc-min grid of surface points. The results are shown in Fig. 7. 5. The gravitational effect of the reference topography varies from $-68. 3$ to 261. 7 mGal. It is mostly negative at low elevations while positive at higher elevations (see Fig. 7. 3). The gravitational contribution of the anomalous topographic density varies between $-24. 0$ and 21. 7 mGal. The large areas of positive values are in the central Otago region. Elsewhere within New Zealand, this effect is mostly negative due to the rock densities which are mainly below $2,670$ kg m^{-3}.

The gravitational contributions of the average and anomalous topographic density distribution to the geoid-to-quasigeoid correction were computed on a 1×1 arc-min grid. The results are shown in Fig. 7. 6. The effect of the reference topography varies from $-105. 3$ to 0. 0 cm, and the effect of the anomalous topographic density is between $-4. 5$ and 11. 2 cm. The large areas of negative values are in the central Otago region. Elsewhere within New Zealand, the gravitational contribution of the anomalous topographical density is mainly positive with the local maxima in the Southern Alps.

The application of the average density evaluated specifically for each island can provide more realistic estimates of the topographic effect in case of considering only a uniform density model. Significantly more realistic estimates of the topographic

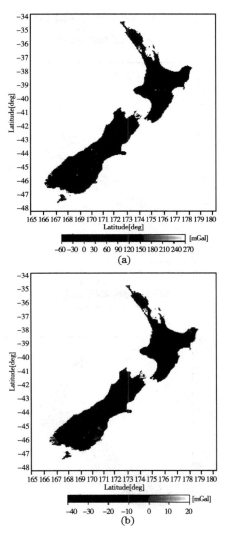

Fig. 7. 5   The topographic effect on gravity of: (a) the reference density
$(2,670 \ kg \ m^{-3})$, and(b) the anomalous density.

effect are, however, expected when utilizing the rock density
model. The results revealed that the gravitational contributions of
the anomalous topographic density on the gravity (varying

between −24. 0 and 21. 7 mGal) and geoid (varying between −9. 5 and 11. 2 cm) represent ~15% of the total topographic effect.

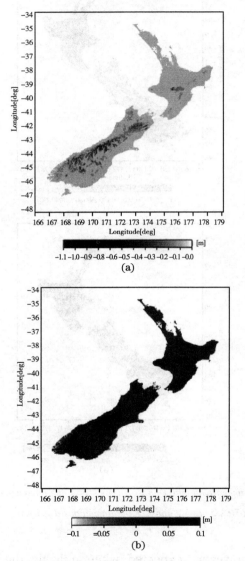

Fig. 7. 6 *The topographic effect on the geoid-to-quasigeoid correction of:* (a) *the reference density* (2,670 kg m⁻³), *and* (b) *the anomalous density.*

8. Conclusions

It is a well-known fact that Molodensky's concept does not require any hypothesis on the density distribution within the topography. The normal heights can then be computed with the accuracy, limited only by the cumulative and random errors in geodetic spirit levelling and gravity measurements along levelling lines. Molodensky's concept has also some practical advantages related to the definition of height anomaly for the external gravity field (i. e. the definition of the disturbing potential at the topographic surface) which can be facilitated, for instance, in testing the accuracy of global geopotential models using the GPS and levelling data. Compared to Molodensky's normal heights, Helmert's orthometric heights approximate more closely the actual length of the plumbline between the geoid and topographic surface. This is due to the fact that the Poincaré-Prey gravity gradient in Helmert's definition of the mean gravity satisfies Poisson's equation inside the topography, whereas the normal gravity gradient in Molodensky's definition does not (it assumes no masses above the ellipsoid surface). In Helmert's definition, however, only a uniform topographic density model is assumed, disregarding the effects of terrain geometry and variable topographic density to the gravity gradient within the topography. Moreover, mass density heterogeneities distributed below the geoid surface are not taken into consideration. In Niethammer and Mader's definitions, the gravity gradient within the

101

topography is computed more realistically by incorporating the planar terrain correction into Poincare-Prey's gravity gradient (see also Santos et al. , 2006). In more recent studies, not only the terrain effect, but eventually also the anomalous topographic density distribution was taken into consideration in theoretical derivations and practical computations (e. g. , Vaníček et al. , 1995; Allister and Featherstone, 2001; Tenzer and Vaníček, 2003; Flury and Rummel, 2009; Sjöberg, 2010, 2012). In addition to these effects, Tenzer et al. (2005, 2006) and Sjöberg (2006) took the mass density heterogeneities below the geoid surface into consideration in their accurate method of computing the geoid-to-quasigeoid correction.

Following a theoretical concept of the rigorous orthometric height definition, the numerical procedure of computing the geoid-to-quasigeoid correction from the observed gravity data was modified by means of applying the inverse solution to discretized Green's integral equations for converting the no-topography gravity disturbances/anomalies at the topographic surface to the respective disturbing potential values at the geoid surface. Consequently, the no-topography disturbing potential difference was computed by means of solving the Poisson's discretized integral. A theoretical foundation for this modification was given by Novák (2003), who firstly applied this method in the gravimetric geoid determination. Later, Tenzer et al. (2009) derived the spectral expressions for computing the far-zone contributions to Green's integrals by means of Molodensky's truncation coefficients. According to these numerical procedures, the observed gravity data at the topographic surface (and corrected for the topographic effect) are directly converted to the disturbing potential values at the geoid surface. The gravity-to-potential conversion is thus explicitly incorporated in the downward continuation procedure while in previous studies this conversion was realized in the successive step of the upward continuation procedure. In particular, Tenzer et al. (2005) applied the inverse solution to discretized Poisson's integral equations in a rigorous determination of the orthometric height (and consequently also the geoid-to-quasigeoid correction) for the downward continuation of the no-topography gravity disturbances. They then solved the integral mean of discretized Poisson's integral for computing the non-topographic part of the disturbing potential difference. Tenzer et al. (2006) developed an alternative method of computing the geoid-to-quasigeoid correction from the observed gravity anomalies. They applied the inverse solution to discretized Poisson's integral equations for the downward continuation of the no-topography gravity anomalies and subsequently solved the generalized Stokes problem to compute the no-topography disturbing potential difference from the respective gravity anomalies at the geoid surface (see also Sjöberg, 2006).

The spectral expressions presented in Chapter 5 can be applied to evaluate the long-to-medium wavelength part of the geoid-to-quasigeoid correction in the remove-compute-restore numerical scheme, while the remaining higher-frequency part of this correction could be computed in the spatial domain. According to this spectral approach, the computation of the topographic potential difference is realized individually for the reference and anomalous topographic density distributions. By analogy with Sjöberg (2007), the potential difference of the reference topographic density was defined by means of applying the analytically-continued topographic potential and the topographic bias. The anomalous density distribution within the topography was described by the volumetric mass density contrast layers. The expressions for the gravitational contribution of the volumetric mass density layer (Tenzer et al. , 2012b) were utilized for computing the geoid-to-quasigeoid correction. These expressions, modified for computing the potential difference of the volumetric mass density contrast layer, were defined in terms of the potential coefficients for the external and internal convergence domains. The non-topographic part of the gravity field is then generated using Stokes' coefficients, corrected for the topographic potential coefficients.

Different height systems were adopted for a realization of geodetic vertical datums around the word. Moreover, each country has different requirements regarding the choice of height system specifications. The conversion between different height systems is thus indispensible for the vertical datum unification. The example of the experimental unification of the LVDs in New Zealand was summarized in Chapter 6. For this purpose, the normal-orthometric heights were converted to the normal and orthometric heights. The normal heights at the GPS-levelling points were then used to estimate the LVD offsets based on applying the geopotential-value approach. Both, the orthometric and normal heights were also used for the comparison of the newly adjusted levelling networks with the regional gravimetric geoid and quasigeoid models.

As already emphasized, the effects of terrain geometry, variable topographic density, and mass density heterogeneities distributed below the geoid surface have to be taken into consideration for an accurate determination of the geoid-to-quasigeoid correction, especially in the mountainous, polar, and geologically complex regions. Flury and Rummel (2009), for instance, demonstrated that the effect of terrain geometry significantly reduces the values of the geoid-to-quasigeoid correction computed using the classical definition in which the topography is approximated by the Bouguer plate. In polar regions of Antarctica and Greenland with large thickness of continental ice sheets (reaching maxima up to about 4 km) the ice-thickness data and the density of the glacial ice have to be used to compute the effect of ice to the geoid-to-quasigeoid correction. The effect of lakes is, on the other hand, likely much less significant. Martinec et al. (1995), for instance, estimated that the effect of water density contrast on the geoid (and consequently to the geoid-to-quasigeoid correction) over the lake Superior reaches maxima of only 0. 24 cm. The effect of variable topographic density to the geoid-to-quasigeoid correction, however, is considerably larger. Tenzer et al. (2005) estimated, based on the numerical results in a high-elevation and rugged part of the Canadian Rocky Mountains, that the effect of variable topographic density is within -7 to 2 cm. Another numerical example from New Zealand (in Chapter 7) revealed that the effect of variable topographic density (with respect to the reference density of $2,670$ kg m^{-3}) represents roughly 15% of the total topographic effect. Even the application of a uniform density model can considerably improve the accuracy of the geoid-to-quasigeoid correction. This is evident from the analysis of New Zealand's geological data which revealed that the average densities at the North and South Islands are $2,336$ and $2,514$ kg m^{-3} respectively. These values are significantly smaller than the average density of $2,670$ kg m^{-3}. This corresponds to relative errors of about 12% and 6% respectively in computed values of the geoid-to-quasigeoid correction at these two islands. Tenzer et al. (2005) also investigated the effect of mass density heterogeneities below the geoid surface. They reported the values of this effect within a range of -8 to 44 cm. Since this effect is mostly positive while the total topographic effect (including the variable topographic density) is mostly negative (– 86. 5 to 0. 1 cm), their combined contribution to the geoid-to-quasigeoid correction (–0. 1 to –45. 6 cm) is only about a half of the total topographic effect. These results indicate that the geoid-to-quasigeoid correction could be computed accurately only if all these effects are consistently taken into consideration.

References

Abdalla A. , Tenzer R. , 2011. The evaluation of the New Zealand's geoid model using the KTH method. Geodesy and Cartography, 37(1): 5-14.

Abdalla A. , Tenzer R. , 2012. Compilation of the regional quasigeoid model for New Zealand using the discretized integral-equation approach. Journal of Geodetic Sciences, 2(3): 206-215.

Andersen O. B. , 2010. The DTU10 gravity field and mean sea surface. Second international symposium of the gravity field of the Earth (IGFS2), Fairbanks, Alaska.

Allaby M. , 1999. A dictionary of Earth sciences. Earth Sciences, Oxford University Press, p. 654.

Allister N. A. , Featherstone W. E. , 2001. Estimation of Helmert orthometric heights using digital barcode levelling, observed gravity and topographic mass-density data over part of Darling Scarp, Western Australia. Geomatics Research Australasia, 75: 25-52.

Amos M. J. , 2007. Quasigeoid modelling in New Zealand to unify multiple local vertical datums. PhD Thesis, Curtin University of Technology, Perth, Australia.

Amos M. J. , Featherstone W. E. , 2009. Unification of New Zealand's local vertical datums: iterative gravimetric quasigeoid computations. Journal of Geodesy, 83: 57-68.

Andrew R. L. , 1995. Porphyry copper-gold deposits of the Southwest Pacific. Mining Engineering, 1: 33-38.

Annen C. , Scaillet B. , 2006. Thermal evolution of leucogranites in extensional faults: implications for Miocene denudation rates in the Himalaya. Geological Society, London, Special Publication, 268, pp. 309-326.

Arafin S. , Singh R. N. , George A. K. , Al-Lazki A. , 2008. Thermoelastic and thermodynamic properties of harzburgite—an upper mantle rock. Physics and Chemistry of Solids, 69(7): 1766-1774.

Ardalan A. A. , Grafarend E. W. , 1999. A first test for W_0 the time variation of W_0 based on three GPS campaigns of the Baltic Sea level project, final results of the Baltic Sea Level 1997 GPS campaign. Report of Finnish Geodetic Institute, 99(4): 93-112.

Arrnienti P. , Innocenti F. , Pareschi M. , Pompilio M. , Rocchi S. , 1991. Crystal population density in not stationary volcanic systems: estimate of olivine growth rate in basalts of Lanzarote (Canary Islands). Mineral Petrology, 44: 181-196.

Artemjev M. E. , Kaban M. K. , Kucherinenko V. A. , Demjanov G. V. , Taranov V. A. , 1994. Subcrustal density in homogeneities of the Northern Eurasia as derived from the gravity data and isostatic models of the lithosphere. Tectonophysics, 240: 248-280.

Årgen J. , 2004. The analytical continuation bias in geoid determination using potential coefficients and terrestrial gravity data. Journal of Geodesy, 78:314-332.

Baeschlin C. F. , 1948. Lehrbuch der Geodäsie. Orell Füssli, Zürich.

Balco G. , Stone J. O. , 2003. Measuring the density of rock, sand, till, etc. UW Cosmogenic Nuclide Laboratory, methods and procedures. Unpublished Report.

Bagherbandi M. , Tenzer R. , 2013. Geoid-to-quasigeoid separation computed using the GRACE/GOCE global geopotential model GOCO02S—a case study of Himalayas, Tibet and central Siberia. Terrestrial, Atmospheric and Oceanic Sciences, 24(1): 59-68.

Becker J. J. , Sandwell D. T. , Smith W. H. F. , Braud J. , Binder B. , Depner J. , Fabre D. , Factor J. , Ingalls S. , Kim S. -H. , Ladner R. , Marks K. , Nelson S. , Pharaoh A. , Sharman G. , Trimmer R. , von Rosenburg J. , Wallace G. , Weatherall P. , 2009. Global Bathymetry and Elevation Data at 30 Arc Seconds Resolution: SRTM30_PLUS, revised for Marine Geodesy.

Blick G. , Crook C. , Grant D. , 2005. Implementation of a semi-dynamic datum for New Zealand. In: Sansò F. (ed.), A Window on the Future of Geodesy, Springer, Berlin, Germany, pp. 38-43.

Bomford G. , 1971. Geodesy (3rd ed.). Oxford University Press,

Oxford.

Bruns H. , 1878. Die Figur der Erde. Publication of Preuss Geodetic Institute, Berlin.

Bullen K. E. , 1966. The bearing of dunite on sub-crustal problems. Bulletin of Volcanology, 29: 307-312.

Burke K. F. , True S. A. , Burša M. , Raděj K. , 1996. Accuracy estimates of geopotential models and global geoids. In: Rapp R. H. , Cazenave A. A. , Nerem R. S. (ed.), Proceedings of Symposium No 116 held in Boulder, CO, USA, July 12, 1995. Springer Verlag, Berlin-Heidelberg, Germany, pp. 50-60.

Burša M. , Radej K. , Šíma Z. , True S. A. , Vatrt V. , 1997. Determination of the Geopotential Scale Factor from TOPEX/POSEIDON Satellite Altimetry. Studia Geophysica et Geodaetica, 41: 203-216.

Burša M. , Kouba J. , Kumar M. , Müller A. , Radej K. , True S. A. , Vatrt V. , Vojtíšková M. , 1999. Geoidal geopotential and world height system. Studia Geophysica et Geodaetica, 43: 327-337.

Burša M. , Kouba J. , Müller A. , Raděj K. , True S. A. , Vatrt V. , Vojtíšková M. , 2001. Determination of geopotential differences between local vertical datums and realization of a World Height System. Stud Geophysica et Geodaetica, 45: 127-132.

Burša M. , Kenyon S. , Kouba J. , Šíma Z. , Vatrt V. , Vítek V. , Vojtíšková M. , 2007. The geopotential value W_0 for specifying the relativistic atomic time scale and a global vertical reference system. Journal of Geodesy, 81(2): 103-110.

Carmichael R. S. , 1982. Handbook of physical properties of rocks. Vol. I, Florida, CRC Press, p. 404.

Claessens S. , Hirt C. , Featherstone W. , Kirby J. , 2009. Computation of a new gravimetric quasigeoid model for New Zealand. Technical report prepared for Land Information New Zealand by Western Australia Centre for Geodesy, Curtin University of Technology, Perth, p. 39.

Claessens S. , Hirt C. , Amos M. J. , Featherstone W. E. , Kirby J. F. , 2011. NZGeoid09 quasigeoid model of New Zealand. Survey Review, 43(319): 2-15.

Clark S. P. , 1966. Handbook of physical constants. Revised Edition, New York, The Geological Society of America, INC, p. 587.

Columbus J. , Sirguey P. , Tenzer R. , 2011. A free, fully assessed 15-m DEM for New Zealand. Survey Quarterly, 66: 16-19.

Christensen N. I. , Mooney W. D. , 1995. Seismic velocity structure and composition of the continental crust: a global view. Journal of Geophysical Research, 100: 9761-9788.

čunderlík R. , Mikula K. , Mojzeš M. , 2008. Numerical solution of the linearized fixed gravimetric boundary-value problem. Journal of Geodesy, 82: 15-29.

čunderlík R. , Mikula K. , 2009. Direct BEM for high-resolution global gravity field modelling. Studia Geophysica et Geodaetica, 54: 219-238.

Dayoub N. , Edwards S. J. , Moore P. , 2012. The Gauss-Listing geopotential value W_0 and its rate from altimetric mean sea level and GRACE. Journal of Geodesy, 86(9): 681-694.

Dennis M. L. , Featherstone W. E. , 2003. Evaluation of orthometric and related height systems using a simulated mountain gravity field. In: Tziavos I. N. (ed.), Gravity and Geoid 2002, Department of Surveying and Geodesy, Aristotle University Thessaloniki, pp. 389-394.

Drewes H. , Dodson A. H. , Fortes L. P. , Sanchez L. , Sandoval P. (eds), 2002. Vertical Reference Systems. IAG Symposia 24, Springer, Berlin, p. 353.

Dyda M. , 1994. Density limits in metapelitic recrystallization. Faculty of Natural Sciences, Comenius University, Mlynska Dolina, Slovakia. Unpublished Report.

Featherstone W. E. , Kuhn M. , 2006. Height systems and vertical datums: a review in the Australian context. Journal of Spatial Sciences, 51(1): 21-42.

Filmer M. S. , Featherstone W. E. , Kuhn M. , 2010. The effect of EGM2008-based normal, normal-orthometric and Helmert orthometric height systems on the Australian levelling network. Journal of Geodesy, 84(8): 501-513.

Flury J. , Rummel R. , 2009. On the geoid-quasigeoid separation in mountain areas. Journal of Geodesy, 83: 829-847.

Gaal G. , Front K. , Aro K. , 1981. Geochemical exploration of a Precambrian Batholith, source of a Cu-W mineralization of the Tourmaline Breccia in Southern Finland, Journal of Geochemical Exploration, 15(1-3): 683-698.

Giacomini A. , Buzzi O. , Renard B. , Giani G. P. , 2009. Experimental studies on fragmentation of rock falls on impact with rock surfaces. International Journal of Rock Mechanics and Mining Sciences, 46: 708-715.

Gibb R. A. , 1968. The densities of Precambrian rocks from northern Manitoba. Canadian Journal of Earth Sciences, 5: 433-438.

Gilliland J. , 1987. A review of the levelling networks of New Zealand. New Zealand Surveyor, 271: 7-15.

Grafarend E. W. , Ardalan A. A. , 1997. W_0: an estimate of the Finnish Height Datum N60, epoch 1993. 4 from twenty-five GPS points of the Baltic Sea level project. Journal of Geodesy, 71 (11): 673-679.

Hall M. , Robin C. , Beate B. , Mothes P. , Monzier M. , 1999. Tungurahua Volcano, Ecuador: structure, eruptive history and hazards. Journal of Volcanology and Geothermal Research, 91: 1-21.

Hammer S. , 1939. Terrain corrections for gravimeter stations. Geophysics, 4: 184-194.

Harkness W. , 1891. Solar parallax and its related constants, including the figure and density of the Earth. Government Printing Office.

Hayford J. F. , 1909. The figure of the Earth and isostasy from measurements in the United States. US Coast and Geodetic Survey.

Hayford J. F. , Bowie W. , 1912. The effect of topography and isostatic compensation upon the intensity of gravity. US Coast and Geodetic Survey, Vol. 10, Special Publication.

Heiskanen W. H. , Moritz H. , 1967. Physical geodesy. W. H. Freeman and Co, San Francisco.

Helmert F. R. , 1884. Die mathematischen und physikalischen Theorien der höheren Geodäsie, Vol. 2, Teubner, Leipzig.

Helmert F. R. , 1890. Die Schwerkraft im Hochgebirge, insbesondere in den Tyroler Alpen. Veröff Königl Preuss Geodetic Institute, No. 1.

Hildreth W. , Lanphere M. , Champion D. , Fierstein J. , 2004. Rhyodacites of Kulshan caldera, North Cascades of Washington, postcaldera lavas that span the Jaramillo. Journal of Volcanology and Geothermal Research, 130: 227-264.

Hinze W. J. , 2003. Bouguer reduction density, why 2. 67?

Geophysics, 68(5): 1559-1560.

Hobson E. W. , 1931. The theory of spherical and ellipsoidal harmonics, Cambridge University Press, Cambridge.

Hofmann-Wellenhof B. , Moritz H. , 2005. Physical geodesy (2nd ed.), Springer, Berlin.

Huang J. , Vaníček P. , Pagiatakis S. D. , Brink W. , 2001. Effect of topographical density on the geoid in the Rocky Mountains. Journal of Geodesy, 74: 805-815.

Hwang C. , Hsiao Y. S. , 2003. Orthometric height corrections from leveling, gravity, density and elevation data: a case study in Taiwan. Journal of Geodesy, 77(5-6): 292-302.

Johnson J. , Lorenz R. , 2000. Thermophysical properties of Alaskan Loess: An analog material for the Martian for the Martian polar layered terrain? Geophysical Research Letters, 27 (17): 2769-2772.

Kao S. P. , Rongshin H. , Ning F. S. , 2000. Results of field test for computing orthometric correction based on measured gravity. Geomatics Research Australasia, 72: 43-60.

Kellogg O. D. , 1929. Foundations of potential theory. Springer, Berlin.

Kimura G. , Ikesawa E. , Ujiie K. , Park J. , Matsumura M. , Hashimoto Y. , 2001. A rock of the seismic front in the subduction zone: Mélange including cataclastic fragment of

115

oceanic crust. Vol. 1, Frontier Research on Earth Evolution.

Krakiwsky E. J. , 1965. Heights. MS Thesis, Department of Geodetic Science and Surveying, Ohio State University, Columbus, p. 157.

Ledersteger K. , 1955. Der Schwereverlauf in den Lotlinien und die Berechnung der wahren Geoidschwere. Publication dedicated to Heiskanen WA, Publication of Finnish Geodetic Institute, No. 46, pp. 109-124.

Ledersteger K. , 1968. Astronomische und Physikalische Geodäsie (Erdmessung). In: Jordan W. , Eggert E. , Kneissl M. (eds.) Handbuch der Vermessungskunde, Vol. V. , Metzler, Stuttgart.

Lemoine F. G. , Kenyon S. C. , Factor J. K. , Trimmer R. G. , Pavlis N. K. , Chinn D. S. , Cox C. M. , Klosko S. M. , Luthcke S. B. , Torrence M. H. , Wang Y. M. , Williamson R. G. , Pavlis E. C. , Rapp R. H. , Olson T. R. , 1998. The development of the joint NASA GSFC and the National Imagery and Mapping Agency (NIMA) geopotential model EGM96, NASA Technical Publication, TP-1998-206861, NASA GSFC.

Lilje M. , (Ed.), 1999. Geodesy and Surveying in the Future— The Importance of Heights. LMV Rep, 3, National Land Survey, Gävle, Sweden, p. 418.

Llambias E. J. , Gordillo C. E. , Badlivy D. , 1977. Scapolite Veins in a Quartz Monzodiorite Stock from Los Molles, Mandoza, Argentina. American Mineralogist.

MacMillan W. D. , 1930. The theory of the potential. Dover, New York.

Mader K. , 1954. Die orthometrische Schwerekorrektion des Präzisions-Nivellements in den Hohen Tauern. Österreichische Zeitschrift für Vermessungswesen, Sonderheft 15.

Marti U. , 2005. Comparison of high precision geoid models in Switzerland. In: Tregonig P. , Rizos C. (eds.) , Dynamic planet. Springer, Berlin.

Martinec Z. , Vaníček P. , 1994. Direct topographical effect of Helmert's condensation for a spherical approximation of the geoid. Manuscripta Geodaetica 19: 257-268.

Martinec Z. , Vaníček P. , Mainville A. , Veronneau M. , 1995. The effect of lake water on geoidal height. Manuscripta Geodaetica, 20: 193-203.

Martinec Z. , 1996. Stability investigations of a discrete downward continuation problem for geoid determination in the Canadian Rocky Mountains. Journal of Geodesy, 70: 805-828.

Martinec Z. , 1998. Boundary value problems for gravimetric determination of a precise geoid. Lecture notes in Earth Sciences, Vol. 73, Springer, Berlin.

Martinkova M. , Pros Z. , Klima K. , Lokajicek T. , Kotkova J. , 2000. Experimentally determined P-wave velocity anisotropy for rocks related to the Western Bohemia seismoactive region. Studia Geophysica et Geodaetica, 44(4): 581-589.

Menemenlis D. , Campin J. M. , Heimbach P. , Hill C. , Lee T. , Nguyen A. , Schod-lok M. , Zhang H. , 2008. ECCO2: High resolution global ocean and sea ice data synthesis. Mercator Ocean Quarterly Newsletter, pp. 13-21.

Mikuška J. , Pašteka R. , Marušiak I. , 2006. Estimation of distant relief effect in gravimetry. Geophysics, 71: J59-J69.

Molodensky M. S. , 1945. Fundamental Problems of Geodetic Gravimetry (in Russian). TRUDY Ts NIIGAIK, 42, Geodezizdat, Moscow.

Molodensky M. S. , 1948. External gravity field and the shape of the Earth surface (in Russian). Izvestia CCCP, Moscow.

Molodensky M. S. , Yeremeev V. F. , Yurkina M. I. , 1960. Methods for Study of the External Gravitational Field and Figure of the Earth. TRUDY Ts NIIGAiK, Vol. 131, Geodezizdat, Moscow. English translation: Israel Program for Scientific Translation, Jerusalem 1962.

Moritz H. , 1980. Advanced Physical Geodesy, Abacus Press, Tunbridge Wells.

Moritz H. , 2000. Geodetic Reference System 1980. Journal of Geodesy, 74: 128-162.

Morrow C. A. , Lockner D. A. , 2001. Hayward fault rocks: porosity, density and strength measurements. Open-file report 01-421, US Geological Survey, p. 28.

118

Mortimer N. , 2005. PETLAB: New Zealand's rock and geoanalytical database. Geological Society of New Zealand Newsletter, 136: 27-31.

Nettleton W. D. , Flach K. W. , Nelson R. E. , 1969. Pedogenic weathering of Tonalite in Southern California. Geoderma, 4(4): 387-402.

Niethammer T. , 1932. Nivellement und Schwere als Mittel zur Berechnung wahrer Meereshöhen. Schweizerische Geodätische Kommission.

Niethammer T. , 1939. Das astronomische Nivellement im Meridian des St Gotthard, Part II, Die berechneten Geoiderhebungen und der Verlauf des Geoidschnittes. Astronomisch-Geodätische Arbeiten in der Schweiz, Vol. 20, Swiss Geodetic Commission.

Nott J. , 2003. Waves, coastal boulder deposits and the importance of the pre-transport setting. Cairns, Australia. Earth and Planetary Science Letters, 210 (1-2): 269-276.

Novák P. , 2000. Evaluation of gravity data for the Stokes-Helmert solution to the geodetic boundary-value problem. Technical Report 207, Department of Geodesy and Geomatics Engineering, University of New Brunswick, Fredericton.

Novák P. , 2003. Geoid determination using one-step integration. Journal of Geodesy, 77: 193-206.

Novák P. , 2010. High resolution constituents of the Earth gravitational field. Surveys in Geophysics, 31(1): 1-21.

Oliveira C. D. , Dall'Agnol R. , Batista Corrêa da Silva J. , Arimateia Costa de Almeida J. , 2008. Gravimetric, radiometric, and magnetic susceptibility study of the Paleoproterozoic Redenc, and Bannach plutons, eastern Amazonian Craton, Brazil: Implications for architecture and zoning of A-type granite. Journal of South American Earth Sciences, 25: 100-115.

Pavlis N. K. , Holmes S. A. , Kenyon S. C. , Factor J. K. , 2008. An Earth Gravitational Model to degree 2160: EGM2008, presented at the 2008 General Assembly of the European Geosciences Union, Vienna, Austria, April 13-18.

Pechinig R, Delius H, Bartetzko A. , 2005. Effect of compositional variations on log responses of igneous and metamorphic rocks. II: acid and intermediate rocks. London, Geological Society, London, Special Publications, pp. 279-300.

Pettijohn F. J. , 1975. Sedimentary Rocks, 3rd edition, Harper and Row, New York, p. 628.

Pizzetti P. , 1911. Sopra il calcolo teorico delle deviazioni del geoide dall` ellissoide. Atti R Accademia delle Scienze di Torino, Vol. 46.

Rao M. V. , Prasanna M. S. , Lakshmi K. J. , Chary K. B. , Vijayakumar N. A. , 2008. Elastic properties of charnockites and associated granitoid gneisses of Kudankulam, Tamil Nadu, India. Current Science, 94(10): 1285-1291.

Rapp R. H. , 1961. The orthometric height, M. S. Thesis, Department of Geodetic Sciences, Ohio State University, Columbus, USA, p. 117.

Russell H. , Pellant Ch. , 1981. Encyclopaedia of rocks, minerals, and gemstones, Thunder by press.

Samalikova M. , 1983. Scanning electron microscopy examples of clay residua from crystalline rocks. Bulletin of Engineering Geology and the Environment, 28(1): 91-102.

Sanchez L. , 2007. Definition and realisation of the SIRGAS vertical reference system within a globally unified height system. In Dynamic Planet: Monitoring and Understanding a Dynamic Planet with Geodetic and Oceanographic Tools, 130, pp. 638-645.

Sansò F. , Vaníček P. , 2006. The orthometric height and the holonomity problem. Journal of Geodesy, 80: 225-232.

Santos M. C. , Vaníček P. , Featherstone W. E. , Kingdon R. , Ellmann A. , Martin B. -A. , Kuhn M. , Tenzer R. , 2006. The relation between rigorous and Helmert's definitions of orthometric heights. Journal of Geodesy, 80: 691-704.

Schaeffer P. , Faugere Y. , Legeais F. J. , Picot N. , 2011. The CNES CLS 2011 Global Mean Sea Surface. OST-ST, San-Diego, October 2011.

Schon J. H. , 1996. Physical properties of rocks: Fundamentals and principles of petrophysics. In: Handbook of geophysical exploration, Section I, Seismic exploration, Vol. 18, Pergamon, Great Britain.

Scharroo R. , 2011. Evaluation of CNES-CLS11 mean sea surface. Technical Report 11-001, Altimetrics LLC.

Sjöberg L. E. , 1984. Least squares modification of Stokes and Vening-Meinesz formulas by accounting for errors of truncation, potential coefficients and gravity data. Department of Geodesy, Report 27, University of Uppsala.

Sjöberg L. E. , 1991. Refined least squares modification of Stokes formula. Manuscripta Geodaetica, 16: 367-375.

Sjöberg L. E. , 1995. On the quasigeoid to geoid separation. Manuscripta Geodaetica, 20(3): 182-192

Sjöberg L. E. , 1999. The IAG approach to the atmospheric geoid correction in Stokes's formula and a new strategy. Journal of Geodesy, 73(7): 362-366.

Sjöberg L. E. , 2001. Topographic and atmospheric corrections of the gravimetric geoid determination with special emphasis of the effects of degrees zero and one. Journal of Geodesy, 75: 283-290.

Sjöberg L. E. , 2003a. A solution to the downward continuation effect on the geoid determined by Stokes formula. Journal of Geodesy, 77: 94-100.

Sjöberg L. E. , 2003b. Improving modified Stokes formula by GOCE data. Bolletino di Geodesia e Scienze Affini, 61(3): 215-

225.

Sjöberg L. E. , 2003c. A computational scheme to model geoid by the modified Stokes formula without gravity reductions. Journal of Geodesy, 74: 255-268.

Sjöberg L. E. , 2003d. A general model of modifying Stokes formula and its least squares solution. Journal of Geodesy, 77: 459-464.

Sjöberg L. E. , 2004. A spherical harmonic representation of the ellipsoidal correction to the modified Stokes formula. Journal of Geodesy, 78(3): 1432-1394.

Sjöberg L. E. , 2006. A refined conversion from normal height to orthometric height. Studia Geophysica et Geodaetica, 50: 595-606.

Sjöberg L. E. , 2007. The topographical bias by analytical continuation in physical geodesy. Journal of Geodesy, 81: 345-350.

Sjöberg L. E. , 2010. A strict formula for geoid-to-quasigeoid separation. Journal of Geodesy, 84: 699-702.

Sjöberg L. E. , 2012. The geoid-to-quasigeoid difference using an arbitrary gravity reduction model. Studia Geophysica et Geodaetica, 56: 929-933.

Sjöberg L. E. , Bagherbandi M. , 2012. Quasigeoid-to-geoid determination by EGM08. Earth Science Informatics, 5: 87-91.

Somigliana C. , 1929. Teoria Generale del Campo Gravitazionale dell'Ellisoide di Rotazione. Memoire della Societa Astronomica Italiana, IV, Milano.

Strang van Hees G. L. , 1992. Practical formulas for the computation of the orthometric and dynamic correction. Zeitschrift für Vermessungswesen, p. 117.

Strange W. E. , 1982. An evaluation of orthometric height accuracy using borehole gravimetry. Bulletin Géodésique, 56: 300-311.

Subrahmanyam C. , Verma R. K. , 1981. Densities and magnetic susceptibilities of Precambrian rocks of different metamorphic grade (Southern Indian Shield). Geophysical Journal, 49: 101-107.

Sünkel H. , 1986. Digital height and density model and its use for the orthometric height and gravity field determination for Austria, Proceedings of International Symposium on the Definition of the Geoid, Florence, Italy, May, pp. 599-604.

Sünkel H. , Bartelme N. , Fuchs H. , Hanafy M. , Schuh W. D. , Wieser M. , 1987. The gravity field in Austria. In: Austrian Geodetic Commission (ed.) . The gravity field in Austria. Geodätische Arbeiten Österreichs für die Intenationale Erdmessung, Neue Folge, Vol. IV. , pp. 47-75.

Tamari S. , Samaniego-Martínez D. , Bonola I. , Bandala E. R. , Ordaz-Chaparro V. , 2005. Particle density of volcanic scoria determined by water pycnometry. Geotechnical Testing Journal, 28(4): 321-327.

Tapley B. D. , Chambers D. P. , Bettadpur S. , Ries J. C. , 2003. Large scale ocean circulation from the GRACE GGM01 Geoid. Geophysical Research Letters, 30: 2163-2167.

Tenzer R. , Vaníček P. , 2003. Correction to Helmert's orthometric height due to actual lateral variation of topographical density. Brazilian Journal of Cartography—Revista Brasileira de Cartografia, 55(2): 44-47.

Tenzer R. , 2004. Discussion of mean gravity along the plumbline. Studia Geophysica et Geodaetica, 48: 309-330.

Tenzer R. , Vaníček P. , Santos M. , Featherstone W. E. , Kuhn M. , 2005. The rigorous determination of orthometric heights. Journal of Geodesy, 79(1-3): 82-92.

Tenzer R. , Moore P. , Novák P. , Kuhn M. , Vaníček P. , 2006. Explicit formula for the geoid-to-quasigeoid separation. Studia Geophysica et Geodaetica, 50: 607-618.

Tenzer R. , Novák P. , 2008. Conditionality of inverse solutions to discretized integral equations in geoid modelling from local gravity data. Studia Geophysica et Geodaetica, 52: 53-70.

Tenzer R, Novák P, Prutkin I, Ellmann A, Vajda P. , 2009. Far-zone contributions to the gravity field quantities by means of Molodensky's truncation coefficients. Studia Geophysica et Geodaetica, 53: 157-167.

125

Tenzer R. , Vatrt V. , Abdalla A. , Dayoub N. , 2011a. Assessment of the LVD offsets for the normal-orthometric heights and different permanent tide systems-a case study of New Zealand. Applied Geomatics, 3(1): 1-8.

Tenzer R. , Vatrt V. , Luzi G. , Abdalla A. , Dayoub N. , 2011b. Combined approach for the unification of levelling networks in New Zealand. Journal of Geodetic Science, 1(4): 324-332.

Tenzer R. , Sirguey P. , Rattenbury M. , Nicolson J. , 2011c. A digital bedrock density map of New Zealand. Computers and Geosciences, 37(8): 1181-1191.

Tenzer R. , Čunderlík R. , Dayoub N. , Abdalla A. , 2012a. Application of the BEM approach for a determination of the regional marine geoid model and the mean dynamic topography in the Southwest Pacific Ocean and Tasman Sea. Journal of Geodetic Science, 2(1): 1-7.

Tenzer R. , Novák P. , Vajda P. , Gladkikh V. , Hamayun, 2012b. Spectral harmonic analysis and synthesis of Earth's crust gravity field. Computational Geosciences, 16(1): 193-207.

Tenzer R. , Gladkikh V. , Vajda P. , Novák P. , 2012c. Spatial and spectral analysis of refined gravity data for modelling the crust-mantle interface and mantle-lithosphere structure. Surveys in Geophysics, 33(5): 817-839.

Tenzer R. , Novák P. , Hamayun, Vajda P. , 2012d. Spectral expressions for modelling the gravitational field of the Earth's crust density structure. Studia Geophysica et Geodaetica, 56(1): 141-152.

Tziavos I. N. , Featherstone W. E. , 2001. First results of using digital density data in gravimetric geoid computation in Australia. In: Sideris M. G. (ed.) Gravity, Geoid and Geodynamics 2000, Springer, Berlin, pp. 335-340.

Vaníček P. , Kleusberg A. , Martinec Z. , Sun W. , Ong P. , Najafi M. , Vajda P. , Harrie L. , Tomášek P. , Horst B. , 1995. Compilation of a precise regional geoid. Final report on research done for the Geodetic Survey Division, Fredericton.

Vaníček P. , Tenzer R. , Sjöberg L. E. , Martinec Z. , Featherstone W. E. , 2005. New views of the spherical Bouguer gravity anomaly. Geophysical Journal International, 159: 460-472.

Vankova V. , Kropacek V. , 1974. Gamma-ray absorption and chemical composition of neovolcanic rock. Studia Geophysica et Geodaetica, 18: 173-175.

Vatrt V. , 1999. Methodology of testing geopotential models specified in different tide systems. Studia Geophysica et Geodaetica, 43: 73-77.

Vermeer M. , 2008. Comment on Sjöberg (2006) " The topographic bias by analytical continuation in physical geodesy". Journal of Geodesy, 81 (5): 345-350. Journal of Geodesy, 82: 445-450.

Wibberley C. , McCaig A. , 2000. Quantifying orthoclase and

albite muscovitisation sequences in fault zones. Chemical Geology, 165: 181-196.

Wirth B. , 1990. Höhensysteme, Schwerepotentiale und Niveauflächen. Geodätisch-Geophysikalische Arbeiten in der Schweiz, Vol. 42, Swiss Geodetic Commission.

Woollard G. P. , 1962. The relation of gravity anomalies to surface elevation, crustal structure, and geology. University of Wisconsin Geophysics and Polar Research Center Research Report, 62, p. 9.

Woollard G. P. , 1966. Regional isostatic relations in the United States. In: Steinhart J. S. , Smith T. J. (eds.), The Earth beneath the continents, American Geophysical Union, Geophysical Monograph, 10: 557-594.